CAMBRIDGE COUNTY

General Editor: F. H. H. GUILLEMARD, M.A., M.D.

HERTFORDSHIRE

Cambridge County Geographies

HERTFORDSHIRE

by

R. LYDEKKER

With Maps, Diagrams and Illustrations

Cambridge:
at the University Press
1909

CAMBRIDGE UNIVERSITY PRESS
Cambridge, New York, Melbourne, Madrid, Cape Town,
Singapore, São Paulo, Delhi, Mexico City

Cambridge University Press
The Edinburgh Building, Cambridge CB2 8RU, UK

Published in the United States of America by Cambridge University Press, New York

www.cambridge.org
Information on this title: www.cambridge.org/9781107669505

© Cambridge University Press 1909

First published 1909
First paperback edition 2013

A catalogue record for this publication is available from the British Library

ISBN 978-1-107-66950-5 Paperback

CONTENTS

CONTENTS

ILLUSTRATIONS

MAPS

The illustrations on pp. 7, 32, 60, 62, 64, 83, 86, 92, 93, 94, 95, 100, 101, 102, 106, 108, 109, 116, 118, 119, 121, 122, 123, 124, 126, 127, 130, 133, 139, and 143 are reproduced from photographs by The Homeland Association, Ltd.; and those on pp. 4, 12, 15, 24, 39, 46, 69, 72, 73, 77, 99, 103, 107, 117, 120, 129, 144, 145, 155, 161, and 165 are from photographs by Messrs F. Frith & Co., Ltd., Reigate. Messrs H. W. Taunt & Co., of Oxford, supplied the views on pp. 21, 84 and 131; Mr A. Elsden, of Hertford, those on pp. 134 and 157; and Mr H. Valentine, of Harpenden, the one on p. 5.

1. County and Shire. The Name *Hertfordshire*. Its Origin and Meaning.

The only true and right way of learning geography (which in its widest sense comprehends almost everything connected with the earth) is to become acquainted with the geography—or, strictly speaking, the topography— and history of the district in which we live. Modern England is split up into a number of main divisions known as *counties*, and in some instances also as *shires*; the word *shire*, when it is used, being added at the end of the county name. Thus we have the county of Essex or the county of Hertford; but whereas in the former case the word *shire* is never added, it may be in the case of the latter. We then have either the county of Hertford, or Hertfordshire, as the full designation of the territorial unit in which we dwell. In official documents our area is always mentioned under the title of the "County of Hertford"; but imperfectly educated persons when filling in such documents frequently write the "County of Hertfordshire." This is wrong and superfluous; shire being equivalent to county.

The word *county* signifies an area of which a *count* or *earl* is the titular head. Here it may be incidentally mentioned that the title " Earl " is of Saxon origin, which it was attempted to replace after the Conquest by the Norman-French " Count " ; the attempt being successful only in the case of an Earl's consort, who is still known as " Countess." It also obtained in the case of " County," which is thus practically equivalent to " Earldom."

It now remains to enquire why some counties are also known as shires, while others are not thus designated. In Anglo-Saxon times England, in place of being one great kingdom, was split up into a number of petty kingdoms, each ruled by a separate sovereign. Essex was then a kingdom by itself, situated in the east of the country ; while Wessex was a western kingdom, and Mercia a sovereignty more in the heart of the country. Essex and Sussex, being small kingdoms, were constituted counties by themselves when the country came under a single dominion, and their names have consequently remained without addition or alteration from Anglo-Saxon times to our own day. The larger kingdoms, such as Wessex and Mercia, were, on the other hand, split up into *shares*, or *shires*,—i.e. that which is shorn or cut off —and their names have disappeared except as items in history. Hertfordshire, then, is in great part a *share* of the ancient kingdom of Mercia, of which, indeed, it seems to have formed a centre, as the Mercian kings spent at least a portion of their time at Berghamstedt (Berk- hampstead). It is however only the larger western portion of the county that belonged to Mercia, a smaller area on the eastern side originally being included in the kingdom of Essex.

As to the meaning of the name Hertford, there has been some difference of opinion among archaeologists. In that extremely ancient chronicle, " Domesday Book," the name, it appears, is spelt Herudsford, which is interpreted as meaning "the red ford." The more general and obvious interpretation is, however, that of "hart's-ford," from the Anglo-Saxon *heort*, a hart, or stag ; and this explanation is supported by the occurrence in other parts of the country of such names as Oxford, Horseford, Gatford (= goat's-ford), Fairford (= sheep's-ford), and Swinford. Writing on this subject in a paper on Hertfordshire place-names published in 1859, the Rev. Henry Hall, after alluding to the custom of naming fords after animals, concluded as follows :—" At all events, the custom is so prevalent, and the word hart so common for Anglo-Saxon localities, as Hart's-bath, Hartlepool (the Hart's pool), Hartly—that though several other derivations have been given for the capital of the county, none seems so simple, or so satisfactory, as that which interprets it to mean the hart's ford."

This interpretation has been adopted by that division of His Majesty's local forces formerly known as the Hartfordshire Militia. Possibly it is supported by the title of a neighbouring village, Hertingfordbury, that is to say, the stronghold near Hertingford,—the ford at the hart's meadow. Whether or not it has anything to do with the matter, it may be worthy of mention that red-deer antlers occur in considerable abundance buried in the peat of Walthamstow, lower down the Lea valley, in Essex.

2. General Characteristics of the County.

Hertford is an inland county, situated in the south-eastern portion of England, and cut off from the nearest sea by the whole width of Essex, which forms the greater

Modern Hertfordshire: Station Road, Letchworth

portion of its eastern border. Neither has it any great river of its own communicating with the ocean ; although the Lea, which is navigable below Hertford, and falls into the Thames at Barking in Essex, affords the means of transporting malt (the great output of Ware) and other products to London and elsewhere by water. As to canal communication, this will be discussed in a later section.

Originally Hertford was essentially an agricultural county, as it is to a great extent at the present day ; its

Ancient Hertfordshire : Thatched Cottages, Harpenden

northern three-quarters being noted for its production of corn. The southern portion, on the other hand, was partly a hay-growing and grazing country. Nowadays, however, more especially on the great lines of railway, conditions have materially altered ; and large areas have become residential districts, which in the more southern part are little more than suburbs of the metropolis. Printing-establishments and factories—moved from London for the sake of cheapness—have likewise been set up on the outskirts of many of the larger towns, such as St Albans and Watford, or even in some of the villages. On the other hand, the old-fashioned timbered and tiled or thatched cottages formerly so characteristic of the county are rapidly vanishing and giving place to the modern abominations in brick and slate. Gone, too, is the old-fashioned and picturesque smock-frock of the labourer and the shepherd, which was still much in evidence some five and forty years ago, or even later; its disappearance being accompanied by the loss of many characteristic local words and phrases, to some of which reference will be made in a later section. The gangs of Irish mowers and reapers which used to perambulate the county at hay and harvest time are likewise a feature of the past.

The scenery of the southern portion of the county differs—owing to its different geological formation—very markedly from that of the northern two-thirds; the latter area representing what may be called typical Hertfordshire. Although there is nothing grand or striking in the scenery of this part, for quiet and picturesque beauty—whether of the village with its ancient church nestling in the shelter of the well-wooded valley, or the winding and tall-hedged lanes (where they have been suffered to remain)—it would be hard to beat; and in many instances is fully equal in charm to the much-vaunted Devonshire scenery, although, it is true, the hedge-banks lack the abundant growth of ferns characteristic of those of the latter county. Very characteristic of this part of the county are its open gorse or heath commons, like those of Harpenden, Gustard Wood, Bower's Heath, and Berkhampstead. From the higher chalk downs on the northern marches of the county extensive views may be obtained over the flats of

An Old Farm-House near Wheathampstead

Bedfordshire and Cambridgeshire; while in like manner the southern range of chalk hills in the neighbourhood of Elstree presents a panoramic view over the low-lying clay plains of the southern portion of the county and Middlesex.

In former days, it may be mentioned in this place,

the inhabitants of most, if not all, of the English counties had nicknames applied to them by their neighbours; "Hertfordshire Hedgehogs" being the designation applied to natives of this county, while their neighbours to the eastward were dubbed "Essex Calves."

3. Size. Shape. Boundaries.

The maximum length of Hertfordshire, along a line running in a south-westerly and north-easterly direction, is about twenty-eight miles; while its greatest breadth, along a line passing near its centre from the neighbourhood of Tring to that of Bishop's Stortford, is very nearly the same. Owing to its extremely irregular outline, the county has, for its size, a very large circumference, measuring approximately 130 miles. Here it should be stated that the ancient area of the county differs somewhat from that of what is now known as the administrative county. According to the census of 1901, the area of the ancient, or geographical county, included 406,157 acres, whereas that of the administrative county is only 404,518 acres[1], or about 630 square miles. The difference in these numbers is due to the fact that certain portions of the old county, such as that part of the parish of Caddington originally included in Hertfordshire, have been transferred to adjacent counties. The figures relating

[1] In the *Report* of the Board of Agriculture published in 1905 the number of acres is given as 402,856; and this is taken as the basis of calculation in section 10 and in the diagrams.

to population, etc. given in the sequel refer to the administrative county.

In size Hertford may be reckoned a medium county, its acreage being rather less than that of Surrey, and about half that of Essex.

To describe the shape of Hertfordshire is almost an impossibility, on account of its extremely irregular contour; but as its two maximum diameters are approximately equal, it may be said to lie in a square, of which the four angles have been cut away to a greater or less extent in a curiously irregular manner.   The reason of this irregular outline, seeing that only the eastern border is formed to any great extent by a river-valley, is very difficult to guess. Where its south-eastern boundary leaves the Lea valley in the neighbourhood of Waltham Abbey, Middlesex gives off from Enfield Chase a kind of peninsula running in a north-westerly direction into Hertfordshire; while, in its turn, Hertfordshire, a short distance to the south, sends a better-marked and irregular peninsula (in which stands Chipping Barnet) jutting far into Middlesex.   In consequence of this interlocking arrangement, a portion of Hertfordshire actually lies to the south of a part of Middlesex, although, as a whole, the former county is due north of the latter.   Another, but narrower, projection runs from the south-western corner of the county in the neighbourhood of Rickmansworth so as to cut off the north-western corner of Middlesex from Buckinghamshire; while a third prominence, in which Tring is situated, is wedged into Buckinghamshire from the western side of the county.   Other minor projections occur on the north-western and northern border, of which the most pronounced is the one north of Baldock, jutting in between Bedfordshire and Cambridgeshire.

As regards boundaries, Hertfordshire is bordered on the east by Essex, by Middlesex on the south, by Buckinghamshire on the south-west, by Bedfordshire on the north-west, and, to a small extent, by Cambridgeshire on the north. From a short distance below the Rye House to Waltham Abbey the boundary between Hertfordshire and Essex is a natural one, formed by the rivers Stort and Lea; but the other boundaries of our county are, for the most part at any rate, purely artificial.

It should be added that these boundaries at the present day do not everywhere accord with those of half a century ago. The parish of Caddington was, for instance, in former days partly in Hertfordshire and partly in Bedfordshire, but under the provisions of the Local Government Act of 1888, confirmed in 1897, the whole of it was included in the latter county. Certain other alterations were made about the same time in the boundary.

4. Surface and General Features.

The contours of a district depend almost entirely upon the nature of its geological formations, and the action of rain, rivers, and frost upon the rocks of which they are composed. These formations in the case of this county are briefly described in a later section. Here it must suffice to state that hard slaty rocks form jagged mountain ranges, while soft limestones like our Hertfordshire chalk weather into rounded dome-like hills and ridges, and heavy clays form flat plains. As the bed-rock, or, as we may say, foundation, of Hertfordshire, is constituted either by chalk or clay, it is the two latter types of scenery that mainly characterise this county. It is only, however, on the north-western and northern borders of the county, as at Ivinghoe in Bucks, or between Sandon and Pirton, that we find typical chalk scenery, where the downs forming the north-easterly continuation of the Chiltern Hills of Buckinghamshire enter our own area. Here we find the rounded downs and hollow combes characteristic of the South and North Downs of the south-eastern counties; and the same absence of woods, except where artificial foresting has been attempted. Some approach to true chalk scenery is likewise shown on the line of hills on the London side of Elstree, where they overlook the clay plain forming this part of northern Middlesex. There is, however, a difference in the scenery of this part as compared with that of the downs to the north in that these hills are partially covered with a capping of clay or gravel, while the beds or layers of which the chalk is composed slope towards the plains at the base of the ridge instead of in the opposite direction (see section on GEOLOGY).

Elsewhere the chalk is covered over to a greater or less degree, alike on the hills and in the valleys, with thick deposits of clay, sand, and gravel. These communicate to the hills and valleys a contour quite different from that of chalk downs ; and in many parts of the county we have a series of more or less nearly parallel lines of low undulating hills separated by wide, open valleys. Examples of this type of scenery are conspicuous in walking from the valley of the Ver at Redbourn across the hill on which Rothamsted stands, then descending into the riverless valley of Harpenden, again crossing the hill to the east of the latter, and then descending into the valley of the Lea near the Great Northern railway station.

A Typical Hertfordshire Village: Much Hadham

This capping of clay permits the growth of forest on the hills, as do the alluvial deposits in the valleys; so that in its well-wooded character the scenery of the greater part of the county is altogether different from that characterising the bare North and South Downs and the Chiltern Hills. Where the chalk comes near the surface there is a marked tendency to the growth of beech-trees, splendid examples of which may be seen in Beechwood Park. Elsewhere in the chalk districts the elm is the commonest timber-tree; although it should be said that this species of tree was originally introduced into England from abroad.

Reference has already been made to the numerous open commons dotted over the chalk area of Hertfordshire. These appear to have been left as open spaces at the time the country was enclosed, owing to the sterility of their soil, which is unsuited for growing good crops of either corn or grass. Many of these commons, as in the neighbourhood of Harpenden, were enclosed some time previous to the battle of Waterloo, when corn in this country was so dear, and every available piece of land capable of growing wheat consequently of great value. It may be presumed that the commons with the best soil were selected for enclosure; but most of such enclosed commons produce inferior crops, partly, it may be, owing to the plan on which they are cultivated. Till twenty years ago or thereabouts all such commons in the writer's own neighbourhood were divided into a number of parallel strips, separated by grass "baulks"; these strips representing the respective shares of the copyholders of the district, who had the right of grazing on the original common. The absence of hedges rendered it necessary that the same kind of crop should be grown each year on every plot. This made it not worth the while of the occupiers to spend money on high cultivation. Of late years many of these enclosed commons have either been built over, or come under a single ownership, thereby obliterating one more interesting page in the history of the county.

To the south of the Elstree range of chalk hills the scenery suddenly changes, and on emerging into Middlesex from the tunnel through these hills on the Midland Railway we enter an extensive grassy plain, characterised by its abundance of oak trees, the scarcity of elms, and the total absence of ash. Parts of this plain, which is continued through Middlesex to London, form the great grass-growing and hay-producing district of Eastern Hertfordshire.

A plain very similar in character to that south of Elstree is entered upon to the northward of Baldock, just after leaving the line of chalk hills, although the greater portion of this northern plain is situated in Cambridge-shire and Bedfordshire. When travelling northward from London by the Great Northern Railway, observant per-sons, after passing through the tunnels traversing the chalk hills north of Hitchin, cannot fail to notice the general flatness of the country as this great plain is entered. Although so like the southern plain in general appearance, this northern plain, as is noticed in the section on Geology, is composed of much older rocks.

The heights of some portions of the chalk area above mean sea-level are, for this part of England, considerable. Thus Great Offley Church is 554 feet, St Peter's Church, St Albans, 402 feet, Stevenage 306 feet, Hitchin Church 216 feet, while a hill near Therfield attains the height of 525 feet. Hastoe Hill near Tring is 709 feet.

Bancroft, Hitchin

5. Watershed. Rivers.

In commencing this section it will be well to devote a few lines to the proper meaning of the frequently misunderstood term "watershed."   It means the line of water-division, or water-parting ; that is to say the line along a range of hills from which the streams flow in one direction on one side, and in the opposite direction on the other.   The ridge-tiles on a roof form an excellent illustration of a watershed.

Every local stream is separated from the one nearest to it by a watershed ; while a river-system, such as that of the Thames, is separated by a much more important watershed from the other river-systems which take their origin near its source.   Such river-systems, each enclosed by a single watershed which it shares with its neighbours, are known as drainage-areas, or catchment-basins.

Practically the whole of Hertfordshire lies in the Thames drainage-area.   At first sight it would be natural to suppose that the summit of the line of chalk downs forming the continuation of the Chiltern Hills in the neighbourhood of Tring and continuing thence to Dunstable would constitute the watershed between two river-systems.   But this is not the case, for the Thames cuts through the Chiltern Hills between Wallingford and Pangbourne, and thus receives the drainage of both the northern and the southern flanks of that range. The watershed formed by the continuation of the Chilterns in the Tring district is therefore one of second-rate importance.

On the other hand, in the chalk hills near Hitchin we have a watershed of first class, or primary rank, for it divides the Thames catchment-area from that of the Great or Bedfordshire Ouse, which flows into the North Sea miles away from the estuary of the Thames. Only four comparatively small streams flowing into the Ouse basin lie for part or the whole of their course within the limits of the county. The first of these is the Pirre, or Purwell, a small brook which rises in the parish of Ippolits, and, after passing Much Wymondley, flows into the Hiz near Ickleford. The Hiz itself rises in a spring at Wellhead, a short distance southward of Hitchin (formerly also called Hiz), and after receiving the Purwell, flows to Ickleford, where it leaves the county, being joined in Bedfordshire by the Ivel, which rises not far from Baldock, passing Biggleswade to join the Ouse. Below Biggleswade it flows through Tempsford, where it unites with the Ouse. The last of the four streams belonging to the Ouse system is the Rhee (a Saxon term signifying a water-course or river), which springs strongly from the chalk a short distance west of Ashwell, and after passing Accrington Bridge and crossing the Ermine Street, eventually falls into the Cam.

The whole of the remaining rivers of the county belong to the Thames catchment-area. With the exception of the Thame, to be mentioned immediately in a separate paragraph, these form two main systems, namely that of the Colne draining the western and that of the Lea the northern and eastern part of the county; the watershed between these two systems running in a north-westerly and south-easterly direction between St Albans and Hatfield, and thence to the north of Chipping Barnet.

The Thame is almost entirely a Buckinghamshire river, but it rises in our county and runs on the north side of the great watershed formed by the continuation of the Chiltern Hills; this watershed constituting a broad, nearly waterless belt separating the catchment-area of the upper Thame from that of the Lea. The Thame itself springs from three heads in the parish of Tring; the first of these rising near the vicarage, the second at a spot called Dundell, and the third in a spring known as Bulbourne. The three become united at New Mill, whence, after passing through Puttenham, the Thame flows by Aylesbury, in Buckinghamshire, and from there continues its course till it eventually joins the Thames near the village of Dorchester, a considerable distance below Oxford.

Of the tributaries of the Colne system, the most important is the Ver (or Verlume), which rises some distance to the east of Cheverell's Green, on the Watling Street, and passes through the village of Flamstead, and thence near the line of the high road to Redbourn, where it is joined by an intermittent stream, or "bourne," known as the Wenmer, or Womer. The latter crosses the road from Harpenden to Redbourn at the foot of a steep descent; and in the old days its appearance as a running stream was believed to forbode a death, or "some extremity of dangerous import." From Red-bourn the Ver continues its course by way of Shafford Mill, at which place it crosses the high road, to St Albans, where it passes between the present city and the ruined walls of ancient Verulam, to which latter it is generally believed to have given its name. Thence its course is continued through the old nunnery of Sopwell (of which it supplied the extensive fish-ponds), and from that point it flows through Park Street to join the Colne near Colney Street; the latter stream giving the name to the united rivers, although the Ver is considerably the larger of the two constituents. The Colne itself rises in the neighbourhood of Tittenhanger, between St Albans and Hatfield, and passes through London Colney on its way to join the Ver. Near Watford the Colne receives an important tributary in the shape of the Gade, which issues from the chalk in the meadows of Great Gaddesden (to which and Gadebridge it gives the name), and after passing through Hemel Hempstead and Nash Mills, is joined at Two Waters by the Bulbourne. The latter rises at Tring very close to the Bulbourne source of the Thame (the watershed being here very narrow), and runs by way of the Frith, Dagnalls, Aldbury Meads, Dudswell Bottom, and North Church to the north-east side of Berkhampstead, where its volume is increased by two important springs, and thence to Two Waters. There is also an intermittent stream known as the Hertfordshire Bourne, which, when running, flows into the Bulbourne about halfway between Berkhampstead and Boxmoor. It is reported—and apparently correctly— to flow on the average once in every seven years; the point from which it flows may be higher up or lower down in the valley according to the amount of water discharged.

Below Two Waters the Gade (as the united stream is now called) passes through Kings Langley, Hunton Bridge, and Cassiobury Park to join the Colne between that park and Watford. After skirting the north side of Moor Park immediately below the last-named town, the Colne receives the Chess (giving the name to Chesham, in Buckinghamshire), which passes through Sarratt in our own county. The Colne then reaches Rickmansworth, where it forms the boundary between Buckinghamshire and Middlesex, and thence flowing by way of Uxbridge and Colnbrook, discharges itself into the Thames above Staines, after a course of about thirty-five miles.

We pass next to the basin of the Lea or Luy, the Logodunum or Logrodunum of the ancient Britons, and the largest river in the county. The Lea itself takes origin in a marsh at Leagrave, or Luigrave, a little north of Luton (= Lea-town), in Bedfordshire, and after passing through Luton Hoo, in that county, where it is expanded into a large artificial lake, enters Hertfordshire at East Hyde, in the valley north of Harpenden, and flows thence by way of Wheathampstead and Brocket (where it again expands into another artificial lake) to Hatfield, whence its course is continued by way of Essendon and the north side of Bayfordbury Park to Hertford. Just before entering the county town the Lea receives the Mimram, or Marran, which rises in the parish of Kings Walden, to the north-west of the Bury, and flows to the eastward of Kimpton Hoo, and thence by way of Codicote, Welwyn, Digswell (with added supplies from local springs), Tewin Water, and Hertingfordbury to its

Netting the Gade, Cassiobury Park

junction with the Lea close to Hertford. In the upper
part of its course the Mimram receives the small brook
known as the Kime, from which Kimpton and Kimpton
Hoo take their names.

In Hertford the Lea is divided into two channels,
one of which runs through the eastern portion of the
town to cross Great Hertham common, while the other,
which is navigable, passes along the western side to join
its fellow at the aforesaid common.

At about one-third the distance between Hertford and
Ware the Lea is largely augmented by the waters of the
Beane (Benefician) and the Rib, which join close to their
union with the main river. The Beane rises from a ditch
in the parish of Ardeley (between Stevenage and Buntingford), and thence flows in a southward direction by way
of Walkern, Aston, Frogmore, Watton, and Stapleford,
passing on its way through Woodhall Park, at the
entrance to which it is reinforced by several strong
springs. Between Watton and Stapleford it also receives
a small brook flowing from Bragbury End.

Starting in an easterly direction from Ardeley or
Walkern, we shall cross the low watershed dividing the
valley of the Beane from that of the Rib; which latter
takes its rise near Reed, and after crossing the Ermine
Street at the south end of Buntingford, flows by way of
West Mill and New Bridge to Braughing, where it is
joined by the Quin, which issues from a spring at
Barkway, and passing by Hormead and Quinbury (to
which it lends its name), reaches Braughing Priory.
Below Braughing the Rib, as it is now called, flows by

way of Hammels, Standon, Barwick, Thundridge, to
cross the Ermine Street at Wade's Mill, and thence *viâ*
Ware Westmill and Ware Park to join the Lea with the
Beane.

From Luton to Hatfield the course of the Lea
pursues a generally south-easterly direction ; at Hatfield
it becomes for a short distance due east, and then trends
to the north-east through Hertford to Ware. Here it
takes a sudden bend, so that the remainder of its course
through the county and the adjacent portion of Essex is
almost due south. A direct line from Luton to the Lea
near Waltham Abbey measures about 40 miles; but,
owing to its great north-easterly bend, the course of the
Lea between these two points is something like 45 miles.

The last two rivers on our list are the Ash and the
Stort, both of which in the upper part of their courses run
from north to south nearly parallel with the Beane and the
Quin. The Ash rises near Brent Pelham and thence
flowing by Much Hadham and through the parish of
Widford, falls into the Lea above Stanstead Abbots.
The Stort, which is the most easterly tributary of the
Lea in the county, is of Essex origin, taking its rise at
the very border of the county, near Meesdon, and
entering it again close to Bishop's Stortford. Some
distance above Stortford it receives one tributary from
the Essex and a second from the Hertfordshire side,
the latter forming for a short distance the boundary
between the two counties. At Stortford the Stort is
wholly within Hertfordshire, but a little below the town
it forms the county boundary for a considerable distance,

Bishop's Stortford (showing the Plain Country)

passing by way of Sawbridgeworth on the Hertfordshire, and Harlow on the Essex side, to join the Lea a short distance below the Rye House. From this point and Hoddesdon nearly to Waltham Abbey the Lea forms the boundary between Hertfordshire and Essex, finally leaving the former county near Waltham. The river now constitutes the line of division between Middlesex and Essex, finally joining the Thames below Bow Bridge, at Barking, after a course of 45 miles.

Here may conveniently be mentioned the celebrated Chadwell spring, near Hertford, which after supplying London with a large amount of water by way of the New River (of which more in a later chapter) for three hundred years, failed temporarily in 1897, so that water began to flow into, instead of out of its basin. Previous to this failure the amount of water discharged daily by the Chadwell spring was estimated at not less than 2,600,000 gallons. In addition to the temporary failure of Chadwell, a spring in Woolmers Park, near Hertford, has of late years completely dried up; both these failures being attributed mainly, if not entirely, to the tapping of the Hertfordshire water-supply by the deep borings in London.

6. Geology and Soil.

By Geology we mean the study of the rocks, and we must at the outset explain that the term *rock* is used by the geologist without any reference to the hardness or compactness of the material to which the name is applied;

thus he speaks of loose sand as a rock equally with a hard substance like granite.

Rocks are of two kinds, (1) those laid down mostly under water, (2) those due to the action of fire.

The first kind may be compared to sheets of paper one over the other. These sheets are called *beds*, and such beds are usually formed of sand (often containing pebbles), mud or clay, and limestone or mixtures of these materials. They are laid down as flat or nearly flat sheets, but may afterwards be tilted as the result of movement of the earth's crust, just as you may tilt sheets of paper, folding them into arches and troughs, by pressing them at either end. Again, we may find the tops of the folds so produced worn away as the result of the wearing action of rivers, glaciers, and sea-waves upon them, as you might cut off the tops of the folds of the paper with a pair of shears. This has happened with the ancient beds forming parts of the earth's crust, and we therefore often find them tilted, with the upper parts removed.

The other kinds of rocks are known as igneous rocks, which have been molten under the action of heat and become solid on cooling. When in the molten state they have been poured out at the surface as the lava of volcanoes, or have been forced into other rocks and cooled in the cracks and other places of weakness. Much material is also thrown out of volcanoes as volcanic ash and dust, and is piled up on the sides of the volcano. Such ashy material may be arranged in beds, so that it partakes to some extent of the qualities of the two great rock groups.

The relations of such beds are of great importance to geologists, for by means of these beds we can classify the rocks according to age. If we take two sheets of paper, and lay one on the top of the other on a table, the upper one has been laid down after the other. Similarly with two beds, the upper is also the newer, and the newer will remain on the top after earth-movements, save in very exceptional cases which need not be regarded here, and for general purposes we may look upon any bed or set of beds resting on any other in our own country as being the newer bed or set.

The movements which affect beds may occur at different times. One set of beds may be laid down flat, then thrown into folds by movement, the tops of the beds worn off, and another set of beds laid down upon the worn surface of the older beds, the edges of which will abut against the oldest of the new set of flatly deposited beds, which latter may in turn undergo disturbance and renewal of their upper portions.

Again, after the formation of the beds many changes may occur in them. They may become hardened, pebble-beds being changed into conglomerates, sands into sandstones, muds and clays into mudstones and shales, soft deposits of lime into limestone, and loose volcanic ashes into exceedingly hard rocks. They may also become cracked, and the cracks are often very regular, running in two directions at right angles one to the other. Such cracks are known as *joints*, and the joints are very important in affecting the physical geography of a district. Then, as the result of great pressure applied sideways, the rocks

	Names of Systems		Characters of Rocks
TERTIARY	Recent & Pleistocene Pliocene		sands, superficial deposits
	Eocene		clays and sands chiefly
SECONDARY	Cretaceous		chalk at top sandstones, mud and clays below
	Jurassic		shales, sandstones and oolitic limestones
	Triassic		red sandstones and marls, gypsum and salt
PRIMARY	Permian		red sandstones & magnesian limestone
	Carboniferous		sandstones, shales and coals at top sandstones in middle limestone and shales below
	Devonian		red sandstones, shales, slates and limestones
	Silurian		sandstones and shales thin limestones
	Ordovician		shales, slates, sandstones and thin limestones
	Cambrian		slates and sandstones
	Pre-Cambrian		sandstones, slates and volcanic rocks

may be so changed that they can be split into thin slabs, which usually, though not necessarily, split along planes standing at high angles to the horizontal. Rocks affected in this way are known as *slates*.

If we could flatten out all the beds of England, and arrange them one over the other and bore a shaft through them, we should see them on the sides of the shaft, the newest appearing at the top and the oldest at the bottom, as shown in the figure. Such a shaft would have a depth of between 10,000 and 20,000 feet. The strata are divided into three great groups called Primary or Palaeozoic, Secondary or Mesozoic, and Tertiary or Cainozoic, and the lowest of the Primary rocks are the oldest rocks of Britain, which form as it were the foundation stones on which the other rocks rest. These may be spoken of as the Pre-Cambrian rocks. The three great groups are divided into minor divisions known as systems. The names of these systems are arranged in order in the table with a very rough indication of their relative importance, though the divisions above the Eocene have their thickness exaggerated, as otherwise they would hardly show in the figure. On the right hand side, the general characters of the rocks of each system are stated.

With these preliminary remarks we may now proceed to a brief account of the geology of the county.

In Hertfordshire, apart from the soil and the superficial accumulations of gravel, sand, and clay, only the lower beds or strata of the Tertiary and the uppermost formations of the Secondary period are represented.

As the greater portion of the subjacent rocks of the

county is formed by the Chalk, it will be convenient to commence with this formation. The Chalk extends, or "strikes," across all but the south-eastern portion of the county in a broad belt, with a general south-westerly or north-easterly direction, reaching on the northern side, with a few exceptions, to the border of the county and beyond, while to the southward its boundary runs approximately through Bushey, South Mims, Hertford, and Bishop's Stortford. At Dunstable the Chalk forms what is called an "escarpment," that is to say a high and somewhat precipitous (although rounded) cliff overlooking the great plain formed by the marls and clays of the underlying strata. As in all true escarpments, the beds, or strata of the Chalk, which are somewhat tilted by earth-movements out of their originally horizontal plane, incline, or "dip" away from the main face of the cliff, that is to say, towards the south-east; and this south-easterly dip of the Chalk, apart from local interruptions and folds, continues to its southerly boundary. Now since the Chalk is a porous formation admirably fitted to collect and retain the rain-water falling upon it, while it is underlain, as we shall see shortly, by the impervious Gault Clay of the Bedford-shire plain, and overlain along its southern boundary by the equally impervious London Clay, it is obvious that it will hold all the water thus collected, and that this water will tend to run deep down in the rock in a south-easterly direction. Hence the northern part of the Chalk zone forms an almost perfect water-collecting area, which can be tapped along the southern side of the county by boring through the overlying London Clay.

The Chalk comprises several main divisions, of which the highest is known as the Upper Chalk, or the Chalk with flints; this when fully developed being about 300 feet thick. It is a soft white limestone traversed by nearly horizontal layers of black, white-coated flint, which have originated by a process of "segregation" in the rock subsequent to its deposition as ooze on the old sea-bed. Usually these layers consist of irregular nodular masses; but there is sometimes a continuous thin layer of scarcely more than half-an-inch in thickness, locally known as "chimney-flint." The south-easterly dip of the Chalk is shown by the layers of flint to be not more, as a rule, than three or four degrees. The Upper Chalk extends from the summits of the hills as far down as Rickmansworth, Watford, Hatfield, and Hertford, thus forming the bed-rock of the greater portion of the county. By the wearing away of the overlying Tertiary strata, a small cone, or "inlier," of Chalk is exposed at Northaw.

Next comes a bed of about four feet thick known as the Chalk-rock. It is a hard cream-coloured rock, containing layers of green-coated nodules, is traversed by numerous vertical joints, and rings to the stroke of the hammer. Owing to its hardness, it resists the action of the weather, and is therefore in evidence at or near the summits of the hills, where it can be traced from close to Berkhampstead Castle by Boxmoor and Apsley, and thence to the south-west of Dunstable, Kensworth, the south of Baldock, and so in a north-easterly direction to Lannock Farm.

Below the Chalk-rock we come to the Middle Chalk,

or Chalk without flints, which may be so much as
350 feet in thickness, and rises in a rather steep slope
or "step" from the underlying beds to be next mentioned.
Flints are few and far between in the Middle Chalk,
which forms the western slope of the Downs at Royston,
as well as beyond the limits of the county at Luton, and

View on the Downs looking towards Wallington from
the Icknield Way

so on to the Chiltern Hills. Fossils are much more
numerous in the Middle than in the Upper Chalk. The
lowest bed of the former is the Melbourn Rock, a hard,
nodular band about 10 feet thick. Next comes the grey
and white Lower Chalk, from 65 to 90 feet thick, after
which we reach the Totternhoe Stone. Although only

six feet in thickness, this Totternhoe Stone, which forms the escarpment of Royston Downs, is of importance as having been largely employed in the construction of churches and other buildings on the northern side of the county. It is a sandy grey limestone, which used to be largely quarried at Totternhoe, with special precautions in drying. It can be traced from Tring by way of Miswell, Marsworth, Pirton, and Radwell to Ashwell.

The Totternhoe Stone really forms the top of the Chalk-marl, which is some 80 feet thick, and consists of buff crumbling marly limestones. It forms a strip of low ground at the base of the Chalk escarpment. At the bottom of the Chalk occurs the so-called coprolite-bed, which contains large quantities of phosphate nodules. Forty years ago these beds were extensively worked between Hitchin and Cambridge for the sake of the nodules.

Only on the northern border of the county, between Hitchin and Baldock, and then merely to a very small extent, are any of the beds underlying the Chalk exposed. These comprise, firstly the Upper Greensand, which is either a sandy marl or a sandstone with green grains, and secondly, a dark blue impervious clay known as the Gault. These formations constitute the plain at the foot of the Chalk hills in Bedfordshire, the scenery of which is very similar to that of the London Clay plain in eastern Hertfordshire and Middlesex.

It is important to add that, at a gradually increasing depth as we proceed south, the Gault underlies the whole of the Hertfordshire Chalk, and renders the latter such an

excellent water-bearing formation. If the Gault be perforated we come upon the Lower Greensand, another
excellent water-bearing stratum, which comes to the
surface in the neighbourhood of Silsoe, in Bedfordshire.

We may now turn to the formations overlying the
Chalk in the southern half of the county. Here it should
be mentioned that all the formations hitherto described
overlie (or underlie) one another in what is termed con-
formable sequence; that is to say, there is no break
between them, but a more or less nearly complete passage
from one to another. Between the Chalk and the
overlying Tertiary formations, there is, on the other
hand, a great break or "unconformity"; the surface of
the Chalk having been worn into a very irregular contour,
above which we pass suddenly to the Tertiary beds,
generally containing at their base a number of rolled chalk
flints. This indicates that before the Tertiary beds were
laid down, the Chalk had become dry land; after which
a portion of it once more subsided beneath the ocean.
The Tertiary beds are in fact formed for the most part
from the *débris*, or wearing away of the old Chalk land.

The lowest Tertiary stratum of eastern and southern
Hertfordshire is known as the Woolwich and Reading
beds. These consist of alternations of bright-coloured
plastic clays and sandy or pebble-beds; their maximum
thickness in the county being about 35 feet. They form
a band extending from Harefield Park to Watford,
and thence to Hatfield and Hertford. Below the
Woolwich and Reading beds we come on the London
Clay, of which the basement bed contains a layer of flint

pebbles, although the remainder of this thick formation is a stiff blue clay, turning brown when exposed to the action of the weather. Originally these Tertiary formations must have extended all over the Chalk of central Hertfordshire, as is demonstrated by the occurrence of patches, or "outliers," of them over a zone of considerable width. Such Tertiary outliers occur at Micklefield Hall, Micklefield Green, Sarratt, Abbot's Langley, Bedmond, Bennet's End, and Leverstock Green, and in the northern, or St Peter's portion of St Albans.

Closely connected with these Tertiary formations is the well-known Hertfordshire pudding-stone; a conglomerate formed of stained flint-pebbles cemented together by a flinty matrix as hard as the pebbles themselves, so that a fracture forms a clean surface traversing both pebbles and cement. This pudding-stone is usually found in the gravels (or washed out of them) in irregular masses, weighing from a few pounds to as many tons. It is stated, however, to occur in its original bedding between Aldenham and Shenley; and the rock evidently represents a hardened zone of the Woolwich and Reading beds. Pudding-stone is found in special abundance at St Albans and again in the neighbourhood of Great Gaddesden. In some St Albans specimens the pebbles are stained black for a considerable thickness by the oxides of iron, while the central core is bright red or orange. Such specimens, when cut and polished, form ornamental stones of great beauty; but, on account of their hardness, the expense of cutting is very heavy.

Except on the higher part of the Chalk Downs, and

very generally along a narrow band half way up the sides
of the valleys, the aforesaid formations are but rarely
exposed in the county at the surface, on account of being
overlaid with superficial deposits of gravel, clay, etc.,
which are of post-Tertiary age, and were deposited for the
most part during the time that man has been an inhabitant
of the world.   These superficial beds are very frequently
termed "drift," on account of a large portion being
formed by ice, at the time that northern Europe was
under the influence of the great glacial period.   Over
most of the chalk area the denuded surface of the Chalk is
covered with a thick layer of stiff clay full of flints, this
layer being formed by the disintegration of the Chalk
itself, the soluble calcareous portion being dissolved and
carried away, while the insoluble flints and clay remain.
Above this layer in the neighbourhood of Hertford,
Barnet, and elsewhere, is a series of gravelly beds assigned
to the middle division of the glacial period; while these in
turn are overlaid locally, as at Bricket Wood, between
St Albans and Watford, by the chalky Boulder-clay, of
upper glacial age, which is there some twenty feet in
thickness.   In other places, as at Harpenden, the hills are
capped by a still greater thickness of clayey deposits,
mingled with flints, resting upon a very irregular surface
of Chalk, which appears to be for the most part of glacial
origin.   Speaking generally, Boulder-clay is characteristic
of the east, and clay with flints and gravel of the western
side of the county.

Half way down the sides of the hills, in the district last
named, the Chalk is more or less completely exposed at the

surface along a narrow zone, below which we come upon deposits of gravel, sand, and clay filling the bottoms of the valleys. At Bowling Alley, Harpenden, these deposits are fully forty feet in thickness. Although they have been supposed to be the result of river action, it is more probable that they are due to rain-wash. Indeed this is practically proved in the case of the valley leading from Harpenden towards No-Man's-Land, where the lower end is blocked by a ridge of gravel, which could not possibly have been formed by river action. The stones in these valley-gravels are of irregular shape, and thus quite different from the rounded pebbles of the gravels of the Woolwich and Reading beds, as seen at St Peter's, St Albans. At Harpenden the uppermost layer of valley-gravel is extremely clean and sharp, generally of a golden yellow colour with blackish veins. Deposits of brick-earth occur locally throughout the Chalk area.

Over the greater part of the county the soil is the result of the decomposition of the foregoing superficial formations; and is consequently in most cases of a stiffer and more clayey character on the hill-tops than in the valleys, where it frequently forms only a bed of a foot, or even less, in thickness above the sharp, running gravel. Everywhere in the Chalk districts the soil contains a vast number of flints; but it is, nevertheless, admirably adapted for corn-growing, and especially for malting-grain; Hertford being one of the four English counties best suited to crops of the latter nature. On many of the unenclosed commons the soil is, however, of a poor and hungry nature, producing various kinds of inferior grass, together with spring-flowering gorse, as on Harpenden Common, or heather, as at Kingsbourn Green, between Harpenden and Luton, and at Gustard Wood, near Wheathampstead. On the higher Chalk Downs near Dunstable and Royston there is little or no soil properly so-called; the short, but sweet and nourishing grass growing on the chalk itself. A very different type of soil obtains in the London Clay area in the south and east of the county; this being heavy and clayey, and thus better suited for grass than for corn; in fact in the old days the Middlesex portion of this district was known to the country people as the "Hay-country." Along many of the river-valleys peaty soils of a marshy and swampy nature prevail.

7. Natural History.

In former days, when the mammoth or hairy elephant, the extinct woolly rhinoceros, and the wild ox, together with the African hippopotamus and spotted hyaena roamed over the Thames valley and afforded sport to our pre-historic ancestors, England was joined to the Continent across what is now the English Channel; so that the animals and plants of the southern portion of our islands, at any rate, were more or less nearly identical with those of France and Belgium. The advent of the great ice age, or glacial period, caused, however, a vast disturbance of the fauna and flora (as the assemblages of animals and plants characteristic of different countries are respectively

termed), especially as about this time there occurred
several oscillations in the level of our country, during one
or more of which Great Britain was temporarily separated
from the Continent. How much or how little these and
other changes had to do with the poverty of the British
fauna as compared with that of the Continent is too long
and difficult a question to be discussed in this place ; but

Six Hills, Stevenage (Danish Barrows)

certain it is that even the southern counties of England
do possess fewer species of animals and plants than France
or Belgium ; that this poverty increases with the distance
from the Continent ; and that Ireland is much poorer in
species than England. It may perhaps be well to add,
although it does not really concern our subject, that there
appears to have been another land-connection by means

of which Scotland and Ireland received a portion of their faunas from Scandinavia by way of what is now the North Sea.

At the date of the final insulation of Great Britain from the Continent there is every reason to believe that all the land animals of the former were identical with species inhabiting adjacent regions of the latter. And even at the present day, when isolation has for centuries been exerting its influence on the non-migratory (and in some degree also on the migratory) animals, there are no species of quadrupeds (mammals), birds, or reptiles abso-lutely peculiar to our islands, with the exception of the grouse ; and in the opinion of many naturalists that bird should be regarded rather as a local variety or race of the willow-grouse of Scandinavia than as a distinct species.

Minor variations, however, characterise many, if not indeed all, of our British quadrupeds and birds when contrasted with their continental representatives. The British squirrel is, for instance, very markedly distinct from all the continental races of that animal in the matter of colouring, while somewhat less decided differences characterise our badger, hare, field-mice, etc. Similarly, among birds, the British coal-tit is so decidedly distinct from its continental representatives that it is regarded by some naturalists as entitled to rank as a species by itself ; and minor differences from their continental cousins are displayed by the British redbreast, bullfinch, great titmouse, and many other resident species. Indeed, if careful com-parisons were instituted between sufficiently large series of specimens, it is almost certain that all species of resident

British land animals, as well as exclusively fresh-water fishes, would display certain differences from their foreign representatives; while in some instances, at any rate, as is already known to be the case in regard to certain species, more than one local race of an animal may exist in our own islands.

It is, however, a question as to what is to be gained by the recognition of such comparatively trifling local differences in animals (and still more by assigning to them distinct technical names), as the splitting process may be carried to an almost endless degree. A large London fishmonger is, for instance (as the writer is informed), able to distinguish a Tay from a Severn or Avon salmon; while a wholesale game-dealer will in like manner discriminate between a Perthshire and a Yorkshire grouse. In like manner a Hertfordshire badger or stoat may be distinguished from their Midland or North of England representatives; but it is difficult to see in what respect we should be the better for the recognising of the existence of such differences.

Accordingly, the fauna and flora of Hertfordshire may be regarded for all practical purposes as more or less completely identical with that of the south-east of England generally; and nothing would be gained, even if space permitted, by giving lists of the species which have been found within the limits of our county.

The fauna and flora of Hertfordshire, like those of other counties with varying geological formations, are not, however, by any means the same everywhere. On the contrary, there are well-marked local differences mainly associated with what naturalists call "station"; that is to say, differences of elevation, soil, geological formation, climate, etc., etc. The animals and plants of the high chalk downs in the neighbourhood of Gaddesden, Dunstable, and Ashwell, are for instance more or less markedly distinct from those of the lower level corn-growing areas of the centre of the county. On these elevated tracts we find, for example, wheatears, stone-curlews (near Tring), blue butterflies, burnet-moths, small brown-banded white snails, a periwinkle-like snail with a horny door to its shell known as Cyclostoma, blue gentians, certain orchids, and many other kinds of plants rarely or never seen on the low grounds. The open commons and heaths, on the other hand, as has been already mentioned, are the home of heather and gorse, together with various distinctive birds and reptiles, such as stonechats, whinchats, titlarks, goldfinches, vipers, slow-worms, and lizards. In the river-bottoms and other swampy localities we find marsh and water-birds, such as yellow wagtails, snipes, sandpipers, grebes (at Tring), herons, moorhens, waterrails, coots, dabchicks, and wild duck, together with (locally) the common grass or water snake, amber-snails, marsh-marigolds, purple loose-strife, ragged robin, reeds, and yellow flags.

Beech trees, as mentioned above, form the predominant timber on the chalk-lands other than the high downs, while on the heavier soils of the centre of the county their place is mainly taken by elm and ash. On these lowlands and other open cultivated tracts are found such birds as partridges, corncrakes, lapwings, pipits, and larks;

while in the coppices, hedgerows, and gardens we look for nightingales (from which bird Harpenden takes its name, *haerpen* being a nightingale and *dene* a valley in Anglo-Saxon), blackcaps, whitethroats, wrens, and nuthatches; while the woods are the resort of green and spotted woodpeckers, wood-pigeons, jays, and pheasants. The low grass-growing clay-plains on the southern side of the county support, as already stated, an abundant growth of oaks to the almost complete exclusion of other timber trees; and this area doubtless also presents certain peculiarities in its fauna distinguishing it from the corn-growing tract to the north. The oaks grow to a very great size, especially at Sacombe and Woodhall Park, and three notable specimens in the county are Queen Elizabeth's oak at Hatfield, Goff's oak at Cheshunt, and the Panshanger oak.

In addition to these local peculiarities in the fauna dependent upon elevation, geological formation, soil, and the presence or absence of forest, there are, however, certain others for which climate may possibly account.

A case in point is afforded by the distribution of stag-beetles and magpies in the county. Both these species are unknown in the district immediately round Harpenden, while the former, at any rate, are likewise unknown in the St Albans district, and apparently between that city and London. If, however, we travel from Harpenden to the east, magpies may be met with when we reach Codicote, while in the opposite direction they occur in the Hemel Hempstead district. As to the exact point where stag-beetles make their appearance in the latter direction the writer has no information, but they are to be met with in the neighbourhood of Rickmansworth and elsewhere on the Buckinghamshire border, and become quite common in that county. Grass-snakes, so far as the writer is aware, are likewise absent from the Harpenden neighbourhood, although on the Cambridgeshire side of the county they are quite common, as they are across the border.

If the local distribution of these species were carefully worked out and mapped, we might perhaps be able to account for what is at present a puzzle.

With the increase of population and building the wild fauna of Hertfordshire, like that of England generally, has been gradually becoming poorer in species—probably indeed from the time the mammoth and the woolly rhinoceros were exterminated, as they possibly were, by our prehistoric ancestors. When the wolf, the bear, the wild cat, and the beaver disappeared, is quite unknown ; but it is in comparatively modern times that the marten has been exterminated, a solitary individual of this species having been killed in the county within a score of miles of London, that is to say near Watford, so recently as December, 1872. Polecats appear to have almost if not quite disappeared from the county, although a straggler may occasionally enter from Buckinghamshire, where a few still survive ; and one was trapped in Ware Park about 1885. Otters are rare, although a few occasionally appear in the lower part of the Lea valley, and some may enter the county from Buckinghamshire, in parts of which they are much more common, the Buckinghamshire

Otter-hounds having killed over a score of these animals in 1908. Some years ago badgers were to be found in many parts of the county, a well-known haunt previous to 1840 being "Badger's Dell" in Cassiobury Park. They still occur locally in certain parts of the Buckinghamshire side of the county, and probably elsewhere. Foxes owe their preservation mainly to the sporting instincts of the county gentry and farmers.

Among birds that have disappeared from the county, the most to be regretted is the bustard, which in the early part of last century was still to be found in the neighbourhood of Royston, although the precise date of its extermination from this part of England is unknown. The bittern, too, is, at the very most, known only as an occasional straggler; but a specimen was shot in a small marshy pond on Harpenden Common some time previous to 1860. The Royston crow, by some naturalists held to be only a form of the common crow, has been named from the Hertfordshire town, though a widespread species throughout many parts of Europe.

Neither has extermination been confined to animals. Fern-hunters have in some instances made a clean sweep of certain species of ferns from many districts, if not from the county generally; and nowadays aspleniums, shield-ferns, polypodies, and false maidenhair (*trichomanes*) have completely disappeared from the Harpenden high roads and lanes; the present writer possessing in his garden what he believes to be the sole remaining indigenous specimen of the last-named species.

Among localities specially celebrated for birds in the

county are Tring reservoirs, where vast flocks of water-birds congregate, especially in winter.   Here breeds the great crested grebe ; and here, too, was shot in 1901 the only known British example of the white-eyed pochard. The neighbouring downs, as already mentioned, form one

Tring Park

of the chief English resorts of the stone-curlew, or thick-knee ; a species of especial interest on account of the remarkable manner in which both birds and eggs assimilate to their surroundings.

Rare birds, as well as various maritime species driven from their normal resorts by stress of weather, make their appearance occasionally in various parts of the county, but references to very few of such cases must suffice. During the great visitation of sand-grouse (a bird normally characteristic of the steppes of Central Asia) to the British Isles in 1863, some individuals reached this county. In the early part of last century a little auk, or rotche, was taken on the millhead at Wheathampstead during very severe weather; a great northern diver has been seen on Tring reservoir; a pair of storm petrels were killed some five-and-twenty years ago at Hemel Hempstead, where snow-buntings have likewise been seen; while various species of gulls from time to time put in an appearance in winter. Among recent events of this nature the appearance at Harpenden of an immature specimen of the great purple heron is certainly noteworthy. In 1878 the late Mr J. E. Littleboy had recorded 201 species of birds from the county, and a few others have been added since, bringing up the number to 210 in 1902.

In regard to fishes, it is of interest to quote the following passage from Sir Henry Chauncy's *Historical Antiquities of Hertfordshire*, published early in the eighteenth century. After referring to its other fish, it is there stated, in the author's quaint language, that the river Lea also contains "some Salmons; which (like young Deer) have several denominations: the first Year they are called Salmon-smelts, the second Year Salmon-sprats, the third Year Salmon-forktails, the fourth Year Salmon-peall, the fifth Year Salmonets, and the sixth Year Salmon; and

if these Fish had free Passage by the Mills, and thro' the
Sluices at Waltham up the Stream towards Ware and
Hertford, where they might Spawn in fresh Water and
were carefully preserved from Pochers, they would greatly
increase in that River, and be of great benefit, as well to
the City of London as the Country; for some Water-men
have observed, that they delight in this Stream, and play
much about those Sluices at Waltham."

Chauncy likewise mentions that trout from the Lea
below Hertford, where it has peaty banks, are much less
red than those from the gravelly streams of the chalk
districts.

For botanical purposes the county has been divided
into six districts corresponding to the river-basins; the
first two belonging to the Ouse system, and comprising
the Cam and the Ivel basins, and the other four, comprising the Thame, the Colne, the Brent, and the Lea,
pertaining to the Thames system. Of these the Lea area
is the largest, the Colne next in size, the Ivel considerably
smaller, and the other three quite small. In some respects
the local characters of the flora are, however, better brought
out by taking the geological formations as a basis of
division. The Upper Chalk area, capped with much
Boulder-clay on the eastern, and with clay-and-flints and
gravel on the western side, corresponds very closely as
regards these divisions with the Lea and the Colne basins.
The Middle Chalk, which as we have seen is exposed
on the flanks of the continuation of the Chiltern Hills in
the north-west, is peculiar in being the only area in the
county in which grows the pasque-flower, or anemone;

this chiefly flourishing on south-westerly slopes, as at Aldbury Towers, near Tring. The Middle Chalk is also the chief home of the various kinds of orchis ; the dwarf, the man, and the butterfly orchis being apparently restricted to this formation. The Tertiary area has a vegetation of a totally different type from the so-called " dry-plant type " characteristic of the Chalk area, but it cannot be further mentioned here.

In the five adjacent counties there occur 110 species of flowering plants unknown in Hertfordshire. On the other hand, Hertfordshire has about a dozen plants (exclusive of varieties of the bramble) unknown in the adjacent counties. Of the 893 native flowering plants of Hertfordshire about 110 have not been recorded from Cambridgeshire, while about 120 are wanting in Bedfordshire, 170 in Buckinghamshire, 140 in Middlesex, and 100 in Essex. These figures may, however, be subject to considerable modification by future research. The following passage on the relations of the Hertfordshire flora is quoted from the *Victoria History of the Counties of England* :—

"Taking the number of species in any adjoining county which are absent from Hertfordshire as the best index of the degree of relationship, it would appear that the flora of Bucks is the most nearly allied to that of Herts, and that those of Cambridge and Essex are the most divergent....This is just what might be expected from the physical features and geological structure of these counties. The floras of Cambridge and Essex have also a more northern or north-eastern facies [character] than that of Hertfordshire, which is of a decidedly southern type. The large number of Hertfordshire species which have not yet been recorded from Buckinghamshire is probably due to the flora of that county not having been so thoroughly investigated as ours has been."

8. Climate and Rainfall.

In its original sense the word climate meant the degree of inclination of the sun's rays at any particular spot at a specified date; but nowadays it is employed to designate the average type of weather experienced in a district. In this latter sense it comprises the results of the combined effects of temperature, atmospheric pressure, the degree of moisture in the air, the direction and force of the wind, and the amount of rainfall; the study of climate constituting the science of meteorology. Although the climate of the British Isles is of an exceedingly changeable type, yet the average of the seasonal changes is far from being the same in all parts of the country. Taking temperature alone, we find, for instance, that while the average for the whole year in Shetland is as low as 43° Fahrenheit, in the Scilly Islands it rises to 53°. It is a very general idea that in our islands the winter temperature of a place depends upon whether it is situated in the north or the south. This, however, is a mistake, the temperature having little relation to latitude, but growing colder as we pass from the west to the eastern side of the country; the south of England, as a whole, being milder than the north, not because it is the south, but because it includes such a large extent of land in the west. The degree of elevation above the sea-level has much to do with temperature and the amount of moisture in the air ; high lands being, as a rule, colder and drier in winter than those lying at lower levels. On account of the comparatively high elevation of a considerable proportion of its area and its easterly position, coupled with the prevalence of north-easterly and easterly winds in spring, Hertfordshire ought to have, on the whole, a cold and bracing climate for the greater part of the year ; and, as a matter of fact, such is the case. Indeed, it is commonly said that northern Hertfordshire is so cold and bracing, that only strong and robust constitutions can stand it ; but that for those blessed with such constitutions it is one of the healthiest counties in the kingdom.

Hertfordshire, however, like England on a small scale, has local climates of its own, dependent upon differences in elevation above the sea-level, in the amount of rainfall, in the nature of the geological formation and soil, and also in aspect, especially as regards protection from the east wind in spring. The slope of a hill facing south or south-west receives for instance far more sun in winter than one which looks in the opposite direction ; while it also escapes the full blast of the bitter east wind. Places situated on chalk, and above all on gravel, are drier, and consequently—if on the same level and with a similar aspect—also warmer than those on cold, heavy clay or marl.

As if to confirm and perpetuate the above-mentioned popular error, it happens that the northern and north-

western districts of the county, that is to say, those constituting the chalk area, are very much colder and more
bracing than those to the south and east, whose substra-
tum is clay. This difference depends, however, not on
differences of latitude, but on the lower elevation of the
southern as compared with the northern districts, coupled
with the protection from cold winds afforded to the low
lands by the high ground. To those who are in the habit
of travelling by the Midland railway from the north-
western corner of the county to London, the differences
between the climate of the northern and the southern
districts is made self-apparent in early spring by the extra-
ordinary difference in the condition of the hedges and
trees on the two sides of the Elstree tunnel. It is true
that on this particular route the district south of the
tunnel is in Middlesex, but to the eastward much of
the lowland is in Herts. On the northern side of the
range of chalk-hills pierced by this tunnel the hawthorn
hedges may be seen at a certain period of the spring
to be absolutely devoid of a sign of green; while on
the opposite side they will be in full leaf. There is, in
fact, about a fortnight's difference between the Elstree
and the Mill Hill side of this range in regard to the
development of spring-vegetation; and while the northern
side is exposed to the full force of the east wind, the
combes and valleys with a south-westerly aspect near the
summit of the opposite flank are so warm and sheltered
that hardy species of bamboo and palm will grow in the
open air almost as luxuriantly as in similar situations in
Surrey or Sussex.

We see, then, as has been well remarked, that the division of the county, along the line indicated in the section on its geology, into two very unequal portions—namely, a large north-western area with a relatively dry soil and atmosphere, and a smaller south-eastern tract with a comparatively moist soil and atmosphere—forms a sufficient approach to an accurate climatic division.

Here we may mention that the daily temperature and the amount of moisture in the air, together with the barometric pressure and a number of other details, are recorded at the Meteorological Office in London from reports received from a host of observing stations (either public or private) scattered at intervals all over the country; and at the end of each year the averages, or "means," of these observations are worked out for the British Isles and England generally, and likewise for the various counties and other local districts. This enables comparisons to be instituted between the climates of different places with much greater accuracy than would otherwise be possible.

In the year 1905 the mean temperature for the whole of England was 48·7°, while that of Hertfordshire was 48·9°, as deduced from observations taken at four stations, of which Bennington showed the lowest mean of 48·4°, and New Barnet the highest of 50·2°. The aforesaid county mean of 48·9° was, however, 0·6° above the average ; the average mean for a series of years thus being 47·8°, or about 1° lower than that for England generally.

If we turn to the map here given we notice that, speaking generally, the rainfall of England decreases steadily as we pass from west to east. The moisture-laden clouds, driven by the prevalent winds across the Atlantic, precipitate their contents on reaching the land, more especially if the land be high, and in consequence the country beyond is less wet. Hertfordshire occupies a middle position between the heavy averages of Wales and S.W. England and the minimum of Essex and the neighbourhood of the Wash, as we should expect. The difference between the average of the various stations in our country is remarkable. In the year 1905, which is taken throughout as the basis of comparison, the highest rainfall in England and Wales occurred at Glas Lyn, near Snowdon, and was no less than 176·6 inches; whereas the lowest was registered at Shoeburyness, in Essex, this being only 14·57 inches; while the average rainfall for Great Britain was 27·17 inches. In Hertfordshire, as we shall see, the average in that year for the whole county was 23·47 inches, but this is 1·5 inches below the general average for a series of years, which is 24·52 inches. In the same year the average for the four chief observing stations in the county was, however, 24·22 inches, with a maximum of 25·88 inches at New Barnet and a minimum of 22·51 inches at Bennington. These extremes were exceeded by a maximum of 28·29 inches at Pendley Manor, Tring, and a minimum of 19·31 inches (or rather more than 5 inches above the Essex minimum) at Hillside, Buntingford.

For a succession of years it has, however, been observed that the rainfall of Hertfordshire is in excess of that of all the adjacent counties to the west and south. This is

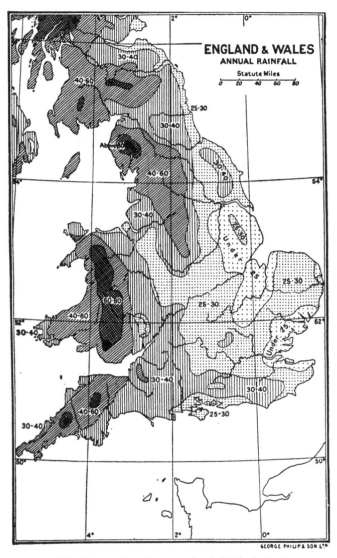

(*The figures show the annual rainfall in inches.*)

shown by the following comparative average rainfalls for 1905; viz.: Bedfordshire (23 stations), 20·47 ins.; Buckinghamshire (32 stations), 22·06 ins.; Middlesex, exclusive of London (49 stations), 22·26 ins.; and Hertfordshire (51 stations), 23·47 ins. Of the 104 stations exclusive of Hertfordshire, the combined mean rainfall is 21·6 inches; Hertfordshire thus showing an excess over the average rainfall in the adjoining counties of nearly two inches in actual amount.

As a whole, and in spite of the excess over its neighbour in the matter of rainfall, Hertford may be reckoned among the relatively dry counties; its average in 1905 being nearly four inches below that for England generally in the same year.

At Bennington there were recorded 1523 hours of bright sunshine during the year, and 54 absolutely sunless days. Throughout Great Britain as a whole there were 186 days in the same year on which a minimum of 0·005 inch of rain fell; all such days with that or a greater quantity of rain being officially known as rain-days. At Greenwich, where the total amount of rainfall was 23·024 inches, the rain-days numbered 161, and the wettest month was June when 4·323 inches of rain were registered. June was also the wettest month of the year in Hertfordshire, but the amount of rain was much less than at Greenwich, being only 3·46 inches.

As regards bright sunshine, the number of hours in England as a whole amounted to 1535, while in Kent the number reached 1667·8, and at Tunbridge Wells 1712·4 hours.

In respect to the number of wet days during the year in question Hertfordshire therefore occupied a very creditable position, although its record for sunshine was less satisfactory.

9. People—Race, Dialect, Settlements, Population.

Previous to the Roman occupation of Britain Hertfordshire was inhabited by two British tribes,—the Cattyeuchlani, whose capital appears to have been Verulam or St Albans, and the Trinobantes. To what extent these original British inhabitants of the county survived the Roman and Saxon invasions is unknown; but it may be taken as certain that at an early date Anglo-Saxon was the language spoken in this part of the country. Forty years ago Anglo-Saxon idioms and words still lingered among the labouring rural population in the Harpenden district (and probably elsewhere), which have apparently now disappeared completely. Instead of *houses*, the Anglo-Saxon plural *housen* was, for instance, always used by the old people; while when a log of timber was cut the wrong way of the grain they would say that it would be sure to *spalt* (equivalent to the German *spalten*), instead of to split.

Of the Anglo-Saxon language there were originally two chief dialects, a northern and a southern; but after the Norman conquest the number of such dialects was increased to half-a-dozen. According to Dr A. J. Ellis's

English Dialects, southern Hertfordshire comes within the domain of the south-eastern dialect, which also prevailed in Middlesex, south-eastern Buckinghamshire, and south-western Essex. Throughout this area there is, however, an underlying basis of the middle eastern dialect, which is still to be detected in northern Hertfordshire, as well as in Essex, Bedfordshire, Huntingdonshire, and Northamptonshire. It accordingly appears that if the East Anglian counties of Cambridgeshire, Suffolk, and Norfolk be eliminated, the whole of the country lying to the eastward of the Chiltern Hills, as well as the high grounds of Northamptonshire, had one dialect in common, which was the speech of the early Teuton settlers of this part of England. This dialect, during the course of fourteen centuries, has been gradually modified and altered by the speech of London till it has resulted in the modern English of this part of south-eastern England.

According to Mr R. A. Smith, writing in the *Victoria History of the Counties of England*, "The grouping of dialects in this part of the country would thus unite Hertfordshire with Essex, and lead us to expect from archæology some indications of Saxon rather than of Anglian influence in the county. The few results already obtained in Hertfordshire certainly show a marked absence of Anglian characteristics, but many discoveries must be made before the peculiarities of East Saxon remains can be demonstrated. To the west of the Chilterns enough has been recovered from graves to show that the settlers in the upper Thames valley, presumably the Saxons of the West, were homogeneous [uniform in characteristics] and distinguishable from their neighbours ; but at present nothing has been found to link them with the people of Essex, who probably reached the eastern slopes of the Chilterns at one time, but were mainly confined to the north of Essex and the neighbourhood of London. In fact, the few discoveries in this district point rather to a connection with Kent [the country of the British tribe of Cantii] than with Wessex."

Whatever may be the precise state of the case with regard to these details, it may be taken as certain that the present population of Hertfordshire is mainly descended from the Saxons who came to Britain in the fifth and sixth centuries. There must have been, however, a certain mixture of ancient British blood ; while later, and more especially among the higher classes, this was followed by the infusion of a Norman strain. Otherwise the population of the county does not appear to have been much influenced by foreign immigration, although there was a settlement of Huguenots in St Albans, who have left their mark in the name of one of the streets—French Row—near the market-place.

Passing on to the present day, we have the somewhat curious anomaly that the population of the administrative county is somewhat larger than that of the original county; this being due to the inclusion in the former of some thickly populated areas. The population of the administrative county in the census of 1901 was given as 258,423 ; ten years previously it was 226,587, thus showing a very marked increase ; that of the ancient county was 250,152 persons. The number of persons to a

French Row, St Albans

square mile in Hertfordshire, on the latter basis at the former date, was thus about 398, against 558 for England and Wales generally. During the last twenty-five years the increase of the population has mainly taken place on the three great lines of railways, the Great Northern, the Midland, and the Great Western, at such places as Barnet, Hitchin, St Albans, Harpenden, and Watford. St Albans has indeed altogether outstripped the county town in point of numbers ; its population being 16,019 in 1901 against the 9,322 of Hertford. In common with England generally, there has of late years been a marked tendency for the rural population to migrate to the larger villages and towns ; and this dominance of the urban population has been accentuated by the transference to St Albans and elsewhere of large manufacturing and printing establishments from the metropolis. The urban districts on the lines of railway likewise constitute the residence of a large population of men having daily business in London.

As in the country generally, the females in 1901 largely exceeded the male population in numbers ; the total for the former being 18,176, and for the latter 16,723. The great majority of these lived in houses, of which 54,963 were inhabited at the date in question. In addition to these, 71 were living in military barracks ; something like another 5,200 were maintained in workhouses, hospitals, asylums, industrial schools, etc., while barges on the canals accounted for about another 147. It should be added that the county contains several large London asylums ; it would therefore give a very exaggerated proportion, as compared with the true state of the case, if the number of lunatics resident in the county were quoted.

Ancient House at Welwyn, now the Police Station

10.   Agriculture—Main   Cultivations, Woodlands, Stock.

As already mentioned, the greater portion of Hertfordshire, that is to say, most of the chalk area, exclusive of the downs, commons, woods, and private parks, was in former years devoted to corn, for the cultivation of which its soil is particularly well suited.   Indeed the county had the reputation of growing not only the best barley for malting, but likewise the best wheat-straw (that is to say, the hardest and whitest) for plaiting. The wheat itself was also of specially good quality and hardness, and there was likewise an abundance of mills in which it could be converted into flour. A noteworthy feature of Hertford-shire agriculture is the practice of mixing chalk with the soils, especially where they are clayey ; this resulting in a decided increase in fertility.

A century and a half ago wheat, barley, and oats formed the chief cereal crops ; beans being better suited to the Vale of Aylesbury, while peas are profitable only on the very light chalky grounds. Clover, lucerne, trefoil, turnips, and (in later times) swedes and mangold are also extensively grown. In this connection it is interesting to note that the first crops of red clover and of swede turnips ever grown in this country were sown at Broadby Farm, near Berkhampstead ; a spot celebrated in literature as having been the home of Peter the Wild Boy in 1725. A certain amount of grass land was intermingled with that under cereal and root cultivation ; while, as mentioned in earlier sections, most of the heavy land in the south and south-east of the county is under grass.

"Hertfordshire farming," observes a recent writer, "has undergone little material change since Ellis's description of it in 1732 ; the hay-crop has become a more prominent feature perhaps, potatoes on the lighter soils have gained a leading place in the rotation, and the standard of fertility has been raised all round ; otherwise a farm on the high chalk-plateau was farmed in 1732 pretty much on the same lines as it is to-day. Ellis gives a list of the chief weeds, 'crow-garlick, wild oat, carlock, poppy, mayweed, bindweed, dock, crow-needle, black bent': they are not less troublesome nor any nearer extinction at the present time, the last grass in particular being very characteristic of corn-land on the 'clay with flints'."

A Hertfordshire Farm near Rickmansworth

Every year the Board of Agriculture publishes a return in which the number of acres in each county devoted to each particular kind of crop is duly recorded, the classification adopted being as follows, viz.: corn crops; green crops; clover, sainfoin, and grasses for hay; grass not for hay; flax; hops; small fruits; and orchards.

Such land as produces none of these crops is classed as bare fallow, of which Hertfordshire in 1905 possessed 14,275 acres.

Of the total of 402,856[1] acres in the county, 329,641 were under cultivation in that year; 1917 were orchards, 26,568 were woodland, while 1657 acres consisted of heaths and commons used as grazing-grounds. At the same date there were 116,700 acres under corn-cultivation; that is to say, something approaching one-fourth the total acreage, against about one-seventh in Kent. Green crops accounted for 32,702 acres, while of the remainder there were 36,831 under clover, sainfoin, and grasses, 3315 under lucerne, meadows claimed 54,589 acres, pasture 70,678, and small fruits 544. Of the corn-grazing area, wheat occupied 51,691, oats 36,946, and barley 27,960 acres.

It is thus apparent that out of the 329,641 acres of cultivated land no less than 200,000, or more than half the whole area of the county, and about 60 per cent. of the total farming land, was still under the plough; this large proportion being at the time exceeded only in six English counties. The increase in permanent pasture has, however, been steadily progressing since the great fall in the price of cereals in the seventies; this being aided by the improvements in the means of communication throughout the country, which have tended to rob Hertfordshire of its original special advantage (owing to its proximity) in the matter of supplying the metropolis with corn and straw.

The subject of Hertfordshire agriculture cannot be

[1] See page 8 and footnote.

dismissed without mention of the fact that the world-renowned agricultural station at Rothamsted, in Harpenden parish, founded and endowed by the late Sir J. B. Lawes, is included within its limits. This includes a laboratory, under a Director, situated on the west side of Harpenden common, and certain plots of land in the park at Rothamsted where agricultural experiments have been carried on for more than sixty years. The whole station is administered by a committee, mainly appointed by the Royal Society.

Fruit is grown only to a comparatively small extent in Hertfordshire. Very characteristic of the county are, however, the orchards (now for the most part more or less neglected) of small black cherries, known as Hertfordshire blacks, and also as " mazzards," which are situated near the homesteads of most of the older farms. These are probably a cultivated variety of the wild black cherry of the neighbouring woods.

On the rich-soiled, low-lying lands of the Lea valley on the south-eastern side of the county are situated numerous market-gardens and nurseries. The growing of tomatos (at Harpenden), cucumbers, and grapes under glass is carried on in several parts of the county on a more or less extensive scale.

Elm, oak, beech, and ash form the most common timber-trees of the county, but the predominance of each kind in particular districts depends, as already mentioned, on the nature of the geological formation. The undergrowth in the woods, which should be cut every 12 or 13 years, consists mainly of hazel.

As regards the number of the larger and commoner kinds of domesticated animals, sheep in 1905 reached a total of 94,461, or about 234 to every 1000 acres; the average for England generally being 445 per 1000 acres. The prevailing breeds are the Hampshire Down, the South Down, and the Dorset; the latter being favoured on account of their early lambing.

Hertfordshire is not a great horse-breeding county, and in 1905 the number of these animals was only 15,070. Cattle numbered 38,636, and pigs 25,338. Shorthorns are the favourite breed of cattle among the farmers; and although in the chalk districts the soil is not specially well suited for dairy purposes, farms near the main railways despatch a considerable amount of milk to London. The number of horses was nearly the same as in 1901, but cattle showed an increase of nearly 2000 head. Sheep, however, had decreased by over 2000 and pigs by more than 6000.

11. Special Cultivations.

The most important special cultivation in Hertfordshire is undoubtedly watercress, which is very extensively grown in the river-valleys over a broad belt of country extending from the Welwyn district, through the parishes of Harpenden, Wheathampstead and Redbourn, and thence to Amersham and Rickmansworth, as well as to the Vale of Aylesbury in Buckinghamshire; this district being reported to be the best in England for this particular crop. The cress is grown in beds cut through the low-ground from one bend of the river to another, so that a constant, but regulated stream of water is continually flowing through. The seed is sown in special beds, and the young cress carefully planted out in regular rows in the mud of the permanent beds during the autumn. Much care has to be exercised in tending and weeding the crop, from which two prolonged cuttings are obtained annually; the spring cutting being, however, much larger and better than the autumn one.

After cutting, the watercress is tied up in bundles and packed in flat, oblong, osier hampers, or baskets ; of which, during the spring season, huge stacks may be seen at the local railway stations awaiting despatch, either to the metropolis, or to the great manufacturing towns of the Midlands. For ordinary purposes the land on which watercress is grown is almost valueless; but the watercress beds yield a big rental. To furnish material for the aforesaid watercress hampers, as well as for basket-work generally, osiers are cultivated in some of the river-valleys, as in the Lea a mile or so above Wheathampstead, where there are extensive beds.

Next in importance to the cultivation of watercress in the Harpenden-Redbourn district is that of lavender at Hitchin, where numerous fields on a spur of the chalk range near Windmill Hill are devoted to the growth of this fragrant plant. In late summer or early autumn the terminal spikes of blossom are nipped from their long stems, and garnered for the sake of their contained oil, which is distilled into lavender-water in Hitchin itself.

A Lavender Field, Hitchin

Lavender-water has been produced at Hitchin for a period of fully eighty years. The growing of lavender as an industry is extremely restricted in England.

12. Industries and Manufactures.

As will be inferred from the statements in earlier chapters with regard to its essentially agricultural nature, Hertfordshire is in no wise a manufacturing county like Lancashire or Yorkshire; this being due, no doubt, in great part to the fact that it possesses no commercially valuable minerals of its own, or, at all events, none which are at present accessible to the miner.

During the last quarter of a century or so a certain number of manufacturing and industrial establishments have been moved from London and set up in various parts of the county, as at St Albans and elsewhere; but these cannot be termed Hertfordshire industries in the proper sense of the term, and do not need further mention.

One of the great industries of the county is the malting business carried on at Ware, as is indicated by the number of cowls over the drying-kilns, which form conspicuous objects for miles round. Ware was the greatest malting place in England. The method of malting is too well known and the industry too widely spread to call for any special notes on the subject.

In former years, say up to about 1865, the straw-plait industry afforded employment to a whole army of workers in north-western Herts and the neighbouring districts of

Bedfordshire and Buckinghamshire; and at that date the women and children might be seen in summer plaiting the straw at almost every cottage-door in each village or town. As already mentioned, the chalk districts of the county grow wheat-straw specially well suited for plait. The straws to be used were selected and pulled one by one from the sheaves before the latter were threshed; and, after having the corn-ears cut off, were done up into bundles. The latter were in turn cut into such lengths as could be obtained free from knots, and tied up into smaller bundles ready for sale to the workers. Before being employed in plaiting, each straw was split longitudinally into several strips by means of a brass instrument, which consisted of a handle and a pointed, star-shaped head bent down at right angles to the former. The finished plait was sold at so much per "score" (that is to say twenty yards) for manufacture into hats and bonnets. Cheap Chinese labour has completely killed the local plaiting industry; but the manufacture of the finished foreign plait into hats still constitutes an important trade in St Albans and elsewhere, as well as at Luton and Dunstable in Bedfordshire; the sewing of the plait into hats being done mainly, if not entirely, in large factories. Tring was a great plait centre.

The manufacture of textile fabrics was formerly carried on in several parts of the county, but in most of these has either completely ceased or fallen into decline. About 1802 there were mills for the manufacture both of silk and cloth at Rickmansworth; and the occurrence of the name "Fuller Street" in the records of St Albans apparently indicates the existence at some unknown date of the latter industry in that city, and there were certainly cotton-mills at Sopwell, to the south of it. The Abbey silk-mills, on the Ver, are still working at St Albans, although with a much diminished output; but those at the neighbouring village of Redbourn have been recently closed. Tring had silk-mills, and canvas was also made

Moor Park, near Rickmansworth

there. Lace-making was probably carried on by some of the cottagers on the Buckinghamshire side of the county, where it flourishes to a certain extent at the present day, and may even still here and there survive.

Nowadays perhaps the most important manufacturing industry in the county is that carried on at the paper-mills at Abbot's Langley, where a large amount of high class paper is turned out.   Being on the canal, these mills have the advantage of water-carriage.   Paper, it may be observed, is made nowadays from wood-pulp and esparto grass, as well as from linen rags.

Brick-making employs a considerable number of hands in various parts of the county ; the glacial and other

Canal and Lock, Rickmansworth

superficial deposits on the chalk area frequently yielding excellent brick-earth ; while the London clay may be employed for brick-making anywhere in the south-eastern districts.   Bricks from the London clay, which is naturally blue, turn yellow or white after burning in consequence of the combustion of the organic colouring matter; but many of those from the glacial clays are of a full rich red.

At Pepperstock, near Caddington, however, there are manufactured certain very hard, heather-coloured bricks, which are much favoured for house-building in north-western Hertfordshire, although their colour compares very unfavourably in the matter of effect when contrasted with the "brick-red" of the more ordinary kinds.

As we approach the Gault plain of Bedfordshire numerous cement works may be seen at the edge of the chalk-marl near the northern borders of the county, some of which may be within the county itself. Chalk is much worked, as at Hitchin and elsewhere, for lime; and, as already mentioned, is dug largely by the farmers for "chalking" their fields. It is to such diggings that the deep circular pits (now generally ploughed over) to be seen in many arable fields are due. The Totternhoe stone has been, and perhaps still is, quarried locally for building in some parts of the northern districts of the county.

13. Minerals—An Exhausted Industry.

Having referred in the last section to brick-making, lime, cement, and Totternhoe stone, very little remains for mention in the present one; as the absence of mines is one of the features of Hertfordshire and the adjacent counties.

Reference may, however, again be made to the so-called coprolite beds of the chalk-marl which were worked in the neighbourhood of Hitchin in the early part of the first half of last century as a source of phosphoric acid for agricultural manure. The irregularly shaped black nodules of phosphate of lime occur crowded together in a comparatively thin bed. They were dug out, and washed from the marl in which they were embedded on the spot in large circular tanks through which a wheel was made to revolve by horse-labour; and then carted away to be converted by a chemical process into superphosphate. After the excavation of the coprolite bed from one strip of a field, the marl from the next was thrown in, and the top soil replaced, so that the land was left in as good, or even better, condition than previously. The industry was continued till all the beds situated at or above a level which it paid to work were exhausted.

In winter there is a considerable local trade in gravel and shingle, dug mainly in the valleys. Twenty years ago this trade was much more extensive in some districts than is at present the case; this being due partly to the exhaustion of the deposits, partly to the fact that in districts served by the Midland Railway the use of syenite from Mount Sorrel and Charnwood Forest, in Leicestershire, has to a considerable extent replaced flint-gravel as road-metal on the main highways of the county and in the metropolitan districts. Formerly, very large quantities of gravel were sent from St Albans and Harpenden to the northern metropolitan suburbs such as Hendon and Child's Hill; but most of that now dug is employed for road-metal on the local by-roads. Here it should be mentioned that in Hertfordshire phraseology the term "gravel" is used exclusively to denote the coarse big-stoned material used as road-metal; what is ordinarily denoted as "gravel"—that is to say the material employed for garden-paths—being locally known as "hoggin." Flints picked from the fields of the chalk area have a higher value as road-metal than dug gravel, owing to their superior hardness; the so-called "quarry-water," which is present in all dug gravel, having been long since dried out.

"Facing" flints for building purposes is an art much less commonly practised in the county than was the case in earlier days; and when buildings of faced flint are contemplated it is generally necessary to send to a distance in order to secure the services of an expert in the facing process.

14. History of Hertfordshire.

The history of Hertfordshire includes such a number of events of primary importance that it is somewhat difficult to make a selection of those most fitted to appear in the limited space available. It was in this county that the offer of the crown of England was made to William the Conqueror, and it was from here that the first petition for the redress of grievances was forwarded to Charles I; while several important battles have been fought within its limits.

To the two British tribes who inhabited this part of England previous to the Roman invasion, reference has been made in an earlier section. The first landing of Julius Caesar took place (in Kent) in 55 B.C., and the second and more successful in 54 B.C.; while a third Roman invasion took place under Claudius in 43 A.D., from which date the Roman legions held possession of the whole country till about the year 410 A.D. Whether Caesar himself ever visited Verulam does not appear to be definitely ascertained, but it was early in the history of that great city that the encounter between the British

The Monastery Gateway, St Albans

Queen Boadicea and the Romans took place. During the Roman period Hertfordshire, which then appears to have been a well-populated and wealthy district, formed a part of the province of Flavia Caesariensis.

The next great event was the Saxon Conquest, which in Kent was ushered in by the landing of a force in the year 449 A.D. During this part of its history the western, or larger portion of our county was included, as already mentioned, in the kingdom of Mercia, while the eastern and smaller section belonged to that of Essex. Of the numerous Mercian kings, the most renowned and most powerful was Offa, whose name survives in Offley, where he had a palace, and where he died about the year 796, while still engaged in building the Monastery and Abbey of St Alban. Mercia at this time made a bid for the supremacy of the petty kingdoms of this part of England, but was eventually beaten by Wessex under the able rule of Egbert.

It was in the reign of the last-mentioned sovereign that invaders of another nationality—namely the Danes—began to make their presence seriously felt in the south; but it was not till the time of his son and successor Aethelwulf that they landed on the east coast. Early in his reign a council of Mercians and West Saxons was held at Kingsbury, near St Albans, to devise means for repelling the invaders; while a second assembly was called for the same purpose at Bennington in the year 850. Neither seems to have resulted in effectual measures, for in 851 we find a large Danish fleet which had sailed up the Thames beating off one of the Saxon kings, who had marched to stop its progress; and after this event the county was harried and raided time after time, till it was eventually divided about the year 880 by a treaty executed at Wedmore between the Saxon sovereign Alfred and Guthrum the Dane by a boundary line running from the mouth of the Lea to its source, and thence straight across country to Bedford. A few years later, however, namely in 894, the Danish fleet sailed up the Lea to Hertford, where Alfred crippled it by cutting into the banks of the river, so that by loss of water the vessels became stranded, and the Danish force had to fight its way to the west of England. After numerous skirmishes and fights, and the building of forts at Hertford and on a small island near Bishop's Stortford, the Danish invasion was practically crushed by King Edward, who died in 925. Much, however, still remained to be done by his son Aethelstan, who stoutly attacked the invaders after they had made a raid on St Albans in 930. A memorial of the Danish sojourn still exists in Dacorum, the name of the western hundred in which Tring is situated; there is also evidence to the same effect in the records of gifts to St Albans Abbey by Danes who had settled in the neighbourhood. The Mercian shire-system, which was probably instituted as an aid against the Danes, is known to have come into force by 957; but in place of Hertfordshire having a sheriff of its own, it shared one with Essex; an arrangement which remained in force till the reign of Elizabeth. This was in Edgar's reign (957—975); but even then we do not reach the end of the Danish trouble, which did not cease till after Sweyn's invasions between the years 1011 and 1014, which were worse than their predecessors, and included the pillage of Canterbury. About this time occurs the first mention of " Heorotford " as the name of the county.

Scarcely had the country recovered, in greater or less degree, from the Danish raids than it was conquered by the Normans under William I, who soon after the battle of Hastings marched through the country south of the

Thames till he reached Berkhampstead in this county, where he built the castle whose foundations and earthworks remain to this day. By relentless severity against all who stood in his way on the march, William had succeeded in instilling a wholesome fear into the Saxon (or, as we now say, English) inhabitants of the country ; and, although he is reported to have been successfully opposed by Frederic, Abbot of St Albans, he was finally tendered the submission of the people and the English crown at Berkhampstead.

To follow in detail the events of the troublous times which succeeded the conquest is here impossible; and it must suffice to state that at Christmas, 1116, Henry I paid a visit to St Albans for the purpose apparently of quelling trouble among the turbulent Norman barons who had now become the paramount lords. Stephen also held a court at St Albans in 1143 in connection with other troubles. With the bare mention that several Hertfordshire barons accompanied Richard I in his crusade to the Holy Land, we may pass on to the quarrel between King John and his barons, which has a very intimate connection with our county ; among the opposing noblemen being Robert Fitzwalter, their leader, and the Earls of Essex and of Hertford. The barons advanced from Northampton to Bedford, while the main body of their army marched to Ware and thence to London. The signing of Magna Charta produced temporary peace ; but this was soon succeeded by fiercer fighting than ever in this county. At the commencement of 1215 John himself was in St Albans, and also had possession of Hertford

and Berkhampstead castles. In May of the same year
Louis landed from France, and in due course besieged
Hertford and Berkhampstead till they surrendered, and
then proceeded to St Albans, where he was for some
time defied by the abbot. At the departure of Louis the
castles were restored to the king.

The next event is the looting of St Albans by
Fulke de Breauté and his band in 1217.

The trouble with the barons continued into the reign
of Henry III; and in the year 1261 the autumn parliament
was held at St Albans. Up to 1295 the shires alone sent
representatives to parliament but in the session held at
St Albans in that year the cities, boroughs, and chief
towns were each permitted to elect two parliamentary
burgesses. During the reign of Edward II the county
was considerably involved in the affairs of Sir Piers de
Gaveston, who spent much of his time at King's Langley,
where Edward had a palace, and where Gaveston was
buried after his execution in 1312. During that year the
papal envoy met the barons at St Albans with a view to
the settlement of their differences with the king; and in
July, 1321, the barons marched through that city on their
way to London. During the fourteenth century the
county suffered severely from plague; but in spite of this
Edward III spent much time at Langley; and in 1361
the king and queen came to Berkhampstead to take leave
of the Black Prince (to whom the castle had been given)
previous to his departure for Aquitaine.

In 1381, owing to exactions on the part of the king
and the abbot of St Albans, there broke out the peasants'

revolt, in which Hertfordshire men took a large share. Indeed after the execution of Wat Tyler the king proposed to go himself to St Albans to punish the insurgents, but was persuaded to send a commission in his stead; although a short time later, after another riot, his majesty appeared in person in that city at the head of an armed force.

With the bare mention that in the second year of his reign King Henry IV visited the abbey, we pass on to the Wars of the Roses, and especially the first battle of St Albans, which was fought in May, 1455. The Lancastrians, or royalists, held the main street of the city till the Yorkists, under the leadership of the Duke of Warwick, burst through the defences from the direction of Sopwell and cut the royalist position in half. In less than an hour they had the city in their own hands, after a great carnage, during which King Henry VI himself was wounded. In 1458 the king visited Berkhampstead with the object of quelling the strife, but to no purpose; and in February, 1461, the two factions again fought an engagement at St Albans, this time at Bernard's Heath, to the northward of St Peter's Church. This second battle of St Albans ended in a victory for the king. On 14th April, 1471, Edward defeated Warwick in the great battle of Barnet, on the south-east border of the county.

Hertfordshire had much to do with royalty during the reign of King Henry VIII, the palace at King's Langley being bestowed on Queen Catherine, while the king himself spent much time at Hunsdon House, and also had a residence at Tittenhanger. There is, moreover, a

The Staircase, Hatfield House

tradition that he was married to Anne Boleyn at Sopwell. The Princess Mary lived for a time at Hertford Castle previous to her removal to Hunsdon, where Princess Elizabeth also lived before her long sojourn at Hatfield, in which beautiful park she was informed of her accession

Cassiobury

to the throne. When queen, Elizabeth continued to be a frequent visitor to the county; and in her reign, owing to plague in London, the law courts were held for a time at St Albans, while, for the same reason, Parliament sat at Hertford in 1564 and 1581. The sovereign herself

came as a guest to Lord Burleigh at Theobalds, to the
Earl of Essex at Cassiobury, and to Sir Nicholas Bacon at
Gorhambury. James I likewise spent much time in the
county, having an establishment at Royston, and dying at
Theobalds. During the civil wars Hertfordshire men
played an important part in connection with what was
known as the Eastern Association; and in 1643, when
the High Sheriff ventured to read a royal proclamation
in the market-place at St Albans, he was arrested by
Cromwell himself. To follow the fortunes of Hertfordshire during the conflict between Charles I and Parliament
would occupy too much space; and it must suffice to
mention that in 1660 Sir Harbottle Grimston of Gorhambury was Speaker of the House of Commons and took a
leading part in the restoration of King Charles II.

The last event in the history of the county to which
space admits allusion is the Rye House Plot. "In the
spring of 1683," to quote the words of a well-known
local writer, "Charles II and James Duke of York went
to see the races at Newmarket. Just opposite to the
Rye House Inn there stood then a castle, built in the
days of Henry VI, and in that castle lived one Rumbold,
formerly an officer in the parliamentary army. Rumbold
and about a score of equally reckless malcontents put their
heads together over their tankards, and, so far as can be
gathered from many rather contradictory narratives, they
formed a plot to delay the royal party on the return
journey from Newmarket to London, by placing an
overturned cart in the road-way, in order that they
might shoot the King and the Duke of York in the

confusion. The conspiracy was frustrated, for the royal party returned earlier than Rumbold had been led to expect, and presently the plot leaked out. The Rye House was searched, incriminating papers were discovered, and the affair culminated in the arrest of those nobler patriots who, in concert with Argyle, had been planning

The Rye House. Portions of the Servants Quarters

the overthrow of what they honestly regarded as a corrupt government."

The Rye House, it may be added, is situated in the south-eastern border of the county, a short distance north-east of Hoddesdon.

15. Antiquities—Prehistoric, Roman, Saxon.

The earliest evidence of the presence of man in Hertfordshire is afforded, as elsewhere in this country, not by written or sculptured records, but by stone implements of various shapes and types. The very earliest of these implements, at any rate, belong to a time when the mammoth inhabited this country, which was then united to the continent; and their age must be reckoned by thousands, if not by tens of thousands, of years. The period to which all these implements belong, being before all human records, is known as the Prehistoric; and it is important to mention that this term should be restricted to the epoch intervening between the time of the formation of the uppermost portion of the Tertiary beds and the first dawn of history. We often find the term "prehistoric monsters" applied to the great reptiles of the Chalk and Oolites; but such a usage, although etymologically justifiable, is technically wrong.

The Prehistoric period for lack of all other means of dating has been divided by antiquarians, according to the material of which man formed his implements, into the Stone, the Bronze, and the late-Celtic or Iron Ages; the Stone age being further divided into an older, or Palaeolithic, section, in which all the so-called "celts," or flint implements, were formed simply by chipping, and a newer Neolithic section, in which they were often ground and polished. In connection with these implements attention may be directed to some of the ancient earthworks in the county, although the age of many of these is unknown, and in some cases may be later than the Roman occupation. The antiquities newer than the late-Celtic age are described as referable to the Roman or the Saxon period, as the case may be. These correspond with the history of England from 55 B.C. to 1066 A.D.

Palaeolithic implements are found locally in certain parts of the county, although from the gravels of a very considerable area, especially the Harpenden district, they appear to be absent. The larger implements, or "celts," which are often six or seven inches in length, seem to have been employed for all purposes, and to have been held in the hand, without handle or shaft, although some of them might easily be used as spear-heads. The first discovery of an implement of this type in the county was made near Bedmond, Abbot's Langley, in 1861. A few specimens have been obtained in other parts of the Colne basin; but in the district round Kensworth and Caddington vast numbers have been discovered, although for the most part just outside the county boundary. In fact, near Caddington the Stone-age men had a great manufactory of these implements :—a kind of Palaeolithic Sheffield. In the basin of the Lea a few flakes, etc., have been found at or near Ayot St Peter, Welwyn, Hertford, Bengeo, Ware, Amwell, Hoddesdon, Ippolits, Stocking Pelham, and elsewhere. Much more important is a "find" at Hitchin, near the source of the Hiz, and thus situated, in part at any rate, in the Ouse basin. These implements, which were first brought to notice in 1877, occur in clay-

Palaeolithic Flint Implement
(*From Kent's Cavern, Torquay*)

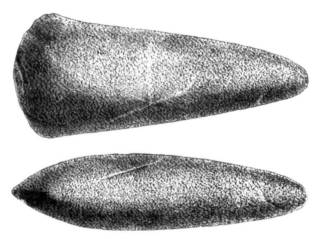

Neolithic Celt of Greenstone
(*From Bridlington, Yorks.*)

pits worked for brick-earth, and are accompanied by bones of the mammoth, hippopotamus, and rhinoceros.

Between the Palaeolithic and Neolithic age exists a gap of untold length, for the land had again to be re-peopled. Chipped, or rough-hewn celts, or hatchets, of the latter age have been picked up in fields near Abbot's Langley, Bedmond, Kensworth, Wheathampstead, Markyate Street, and Weston. Polished celts are more rare, but specimens have been found at Panshanger, King's Langley, Aldbury (near Stortford), Ashwell, and between Hitchin and Pirton. Perforated axe-heads and hammerheads of stone, which may belong to the close of the Neolithic or commencement of the Bronze age, are still more uncommon, although a few such have been found, notably a hammer, near Sandridge, now preserved in the British Museum. Much the same remark applies to chipped arrow-heads—the fairy darts of a more poetical age—but a few beautiful specimens have been found near Tring, some so long ago as the year 1763 or thereabouts, and others at Ashwell and Hunsdon.

After a time man learnt the use of metal. The smelting of iron was at first beyond his power, and he employed the mixture of copper and tin which we term bronze. Of this age specimens of winged celts and palstaves (a narrow hatchet, with a tang or socket for a haft) have been found in various parts of the county, as well as socketed celts, daggers, swords, spear-heads, and the like. The most important discovery of this nature was made in 1876 during drainage operations at Cumberlow Green, near Baldock, when some forty bronze implements were found in a well-like hole. Gold ornaments, probably referable to the same epoch, have been found at Little Amwell and at Mardox, near Ware.

We now come to the early Iron Age, when man had succeeded in mastering this metal. Of this a very brief notice must suffice. Primitive coins, without inscription, of the type issued by Philip II of Macedon, and hence known as Philippi, were probably coined in the county in early British times; but after the Roman invasion a number of British coins were struck at Verulamium, among the most interesting of which are those of Tarciovanus, who reigned in that city from (probably) about 30 B.C. to about 5 A.D. A large number of his coins have been found at Verulam, as well as those of other British sovereigns. Tarciovanus, it may be added, was the father of Cunovelinus (Shakespeare's Cymbeline), whose capital was Camulodunum, the modern Colchester.

Earthworks of great but unknown antiquity are by no means uncommon in Hertfordshire; one of the most important being Grimes-ditch, or Grimm's Dyke, traces of which remain on Berkhampstead common, as well as on the opposite side of the Bulbourne valley, while a deep ditch runs in a bold sweep from near Great Berkhampstead through Northchurch and Wiggington to the north of Cholesbury camp, and thence into Buckinghamshire. Beech Bottom forms another great dyke lying between the site of Verulam and Sandridge, and is probably pre-Roman, and possibly connected with the encampment east of Wheathampstead known as the Moats or the Slad. The latter forms part of a great system of earthworks of which the opposite side is marked by the Devil's Dyke at Marford. The great earthworks running outside of and parallel to parts of the Roman wall at Verulam are likewise older than the latter. Berkhampstead Castle may stand on the site of an earlier camp, as British and Roman coins have been found there ; but the mound or keep, as

The Devil's Dyke, Marford

at Bishop's Stortford, Pirton, and Hertford Castles, is probably Saxon. On the other hand, the well-preserved camp near Redbourn, known as the Aubreys, Auberys, or Aubury, is certainly pre-Roman; and the same is probably the case with some of the numerous other earthworks dotted over the county, such as Arbury Banks, Ashwell. A few barrows or tombs of pre-Roman age have been opened and examined in various parts of the county, as at Therfield, Royston, and Easneye near Ware.

Ancient Causeway, Verulam

In evidence of the Roman occupation of Britain Hertfordshire is unusually rich, although limitations of space prevent anything like justice being done to this part of the subject. The county is, in the first place, traversed

from south to north by three main Roman roads, the
Watling Street, running through St Albans and Markyate,
the Ermine Street, passing through Hertford, and the
Icknield Way, traversing Hatfield and Baldock; as well
as by the Roman Way, connecting the latter town with
Hertford. Of these we shall speak presently. The

Roman Wall in St Germans' Meadow, Verulam

crowning Roman glory of Hertfordshire is, however,
the city of Verulamium, or Verulam, situated on the hill
on the opposite side of the Ver to St Albans. Much of
the foundations of this city lie buried within the area
partially enclosed by the remains of the massive walls;
and the ploughman within that ring is constantly turning
up coins, fragments of pottery and glass, and other articles.

In one place are buried the apparently complete foundations of an amphitheatre, which was opened out many years ago, but again covered up after examination. Of the walls considerable portions, in a more or less damaged condition, still remain to bear eloquent testimony to the lasting character of Roman masonry; and much more would have persisted had they not been used as a

St Albans' Abbey from the South Side

convenient source of materials for the construction of St Albans' Abbey and other ancient buildings. The basement of a Roman villa, in a fine state of preservation, was opened out at Sarratt Bottom in 1908, and plans of the structure prepared, after which the excavations were filled in. Other Roman remains are known to exist in the district.

Ravensburgh Castle, Hexton, is a well-known Roman camp, built on an earlier foundation ; and remains of Roman camps exist at Braughing and several other places in the county, although in many cases the precise age of such ancient stations does not appear to be definitely ascertained.

Isolated Roman remains of various kinds occur in many parts of Hertfordshire. From the writer's own neighbourhood the British Museum possesses a Roman altar found many years ago at Harpenden, as well as a Romano-British stone coffin, containing a glass vessel and pottery, found near Pickford Mill in the Lea valley, east of Harpenden, in 1827. The remains of another Roman interment, including fine specimens of amphorae, or large two-handled pottery vessels, were found about the year 1865 near Harpenden station on the Great Northern railway, and Barkway has yielded a fine bronze statuette of Mars.

Reference may here be made to the ancient mill-stones, for hand use, made of Hertfordshire pudding-stone, and known as querns, of which the writer gave two fine specimens from Harpenden to the British Museum. Both stones have one flat and one convex surface, but the convexity is much greater in the upper stone, which is almost conical, and is completely perforated at the centre. When in use, a stick, to serve as the axis of rotation, was inserted in this hole and received in a socket in the nether stone. The labour involved in making these pudding-stone querns must have been enormous.

With the Saxon period we reach the age of church

building; but apart from such portions of certain churches as are of that age, Hertfordshire is exceedingly poor in evidence of the Saxon dominion. A glass Anglo-Saxon basin, together with a bronze Frankish pot of late sixth or early seventh century work, was, however, discovered at Wheathampstead in 1886. Anglo-Saxon relics are believed also to have been unearthed at Redbourn at a very early period, when they were attributed to St Amphibalus; and a Saxon burial-place appears to have been found near Sandridge in modern times, although unfortunately ploughed over. Apart from the above, there are only a few isolated "finds," such as of the coins known as minimi, and of a gold ornament discovered at Park Street in 1744.

16. Architecture. (a) Ecclesiastical— Abbeys and Churches.

The architecture of Hertfordshire buildings may be most conveniently discussed under three separate sections, namely :—(a) ecclesiastical, or buildings related to the church ; (b) military, or castles ; and (c) domestic, or dwelling houses and cottages.

As in England generally, the architecture of the older buildings of all three classes has been affected to a greater or less degree by the nature of the building materials most easily accessible. Throughout the northern chalk area of the county the Totternhoe stone of Bedfordshire and the northern flanks of Hertfordshire was largely employed in church building, both for inside and outside work, to the latter of which it is but ill suited. Flint—in the better class of work "faced" or "dressed" by fracture so as to present a flattened outer face—was also very extensively used. The Norman builders of the tower of St Albans found, however, a quarry ready to their hands in the adjacent walls of Verulam, and we accordingly find this part of the structure made almost entirely of the characteristic Roman bricks or tiles. Contemporaneous brick was also locally used to a very considerable extent even in the chalk districts; and in the north-western part of the county there are numerous beautiful examples of Tudor brick chimneys, as at Water End. Timber in the old days was, however, much cheaper than bricks, and we consequently find many of the older buildings—especially cottages—constructed of a framework of wood, arranged in the fashion of a net, with the large "meshes" filled in with brick. This type of work is locally known as brick and studding and to the architect as half-timbered work. Other buildings were largely constructed of a wooden framework overlain with lath-and-plaster work.

Many of the churches built of flint or Totternhoe stone have their angles or quoins made of harder material; in many instances of stone apparently from Northampton-shire, but in other cases of Roman brick; similar materials being also used in the arches of some of the churches.

A large number of Hertfordshire churches have relatively low battlemented towers, frequently with a short spire or steeple in the centre as at Tewin, or a turret in one corner. Kensworth is an example of such a battlemented tower without either spire or turret ; St Mary's, Hitchin, Tring, Northchurch, Barnet, Bushey, King's Walden, Cheshunt, and Watford are examples of towers with a turret in one angle, while at Ashwell there are turrets in all four corners. Some of the smaller churches, like St Michael's, St Albans, originally had no

St Peter's, Tewin

aisles. Clothall church is peculiar in that the roof of the tower forms a four-sided cone ; while the roof of the church at Sarratt is equally unique in being saddle-backed, that is to say, having a ridge running at right angles to that of the roof of the nave and chancel.

Apparently there is no wholly Saxon church in the county, although several of the older ones were constructed on the site of Saxon buildings, many of which were probably of wood, and thus either perished through decay or were burnt during the Danish raids.   On the

St Mary's, Cheshunt

other hand, there are remains of Saxon work in St Albans' Abbey, and there are several Hertfordshire churches which are referred with a greater or less degree of certainty to the period before the Norman conquest ; the original part of St Michael's, in St Albans, and of St Stephen's, to the south-west of that town, may be cited as examples, both dating from the middle of the tenth century. The church of Sandridge—on the road from St Albans to Wheathampstead—may likewise date from the same epoch.

St Helen's, Wheathampstead

Of Norman churches there are numerous examples, among which may be cited as a fine specimen St Mary's, Hemel Hempstead, whose tall octagonal tower and spire are visible from a long distance. The Norman arches of the nave are of great solidity, while the western doorway, dating from about 1140, is a magnificent example of the work of the period. Sarratt church is also largely Norman,

as was also the old church of St Nicholas, Harpenden,
unfortunately pulled down (with the exception of the
much later tower) nearly half a century ago. A considerable portion of St Albans' Abbey (now cathedral), the
pride of the whole county, is also Norman; the tower of

St Mary's, Hemel Hempstead

Roman brick being, apart from modern additions, wholly
of that period. The old Saxon Church of King Offa,
which stood on or near the site of the present building,
appears to have been completely swept away by Abbot
Paul of Caen (1077—1093), the founder of the present

St Albans' Abbey

abbey, which although completed by him, was not conse-
crated till 1115. "It is to be inferred," according to the
Victoria History of Hertfordshire, "that a clean sweep was
made of the old buildings, and no evidence as to their site
has been preserved. The Norman Abbot's contempt for
his Saxon predecessors...led him to destroy their tombs,
and he doubtless laid out his new building without
attempting in any way to accommodate them to those
previously existing on the site. But he preserved and
used up in his new church some of the stonework of the
old building, giving a very prominent place to the turned
shafts which still remain in the transept, and are the most
notable relics of the Saxon building." In the present
nave, which is the second longest in England, the first six
pillars on the north side belong to the original structure
of Abbot Paul; after which we come to Early English
(Pointed) work; this being continued to the west end of
the building and back to the fifth pillar on the south side,
whence Decorated work extends to St Cuthbert's screen.
The Norman work (1077—1093) of one side thus faces
Decorated work (1308—1326) on the other, but this is
due to accident rather than design, the Norman pillars
having given way early in the church's history. It has
recently been suggested that the Abbey stands on the site
of the old Roman amphitheatre, and that St Peter's Street,
St Albans, marks the position of the Roman *cursus*, or
race-course.

Of the Decorated style, in vogue during the reigns of
the three Edwards, in other words, throughout the fourteenth century, in addition to the beautiful work in St

Albans' Abbey, we have examples in Abbot's Langley, Clothall, and Hitchin churches. Abbot's Langley has also some fine Norman work in the nave. Many of the churches of the Perpendicular period, like St Mary's, Hitchin, have large and beautiful porches. Most of the windows in Abbot's Langley church are Perpendicular, although some on the south side are Decorated; and Tring and Offley churches are wholly of the Perpendicular style.

Previous to the Reformation, Hertfordshire, like other counties, possessed numerous religious houses, such as priories, monasteries, nunneries, and hospitals; all of which, commencing with the smaller ones, were suppressed by Henry VIII, whose chief agent in the work was Thomas Cromwell. In many instances the sole evidence of the existence of such establishments is the survival of the word "Abbey" or "Priory" as the name of a private mansion, but sometimes their gateways, towers, or merely ruins, still remain.

The neighbourhood of St Albans is especially rich in relics of this nature. To the south, on the banks of the Ver, are the famous ruined walls of Sopwell Nunnery, a building known to have been in existence so early as 1119; but of the monastery there remains only the fine gate-house (long misused as a gaol), together with traces of the cloister arches on the south wall of the abbey. From documentary evidence, however, aided by excavations in the abbey orchard, it has been found possible to make a ground-plan of the whole establishment. The last traces of the hospital of St Mary-de-Pré vanished only during the last century; the name surviving in a private house by the Ver, which is known as the Pré.

Hitchin formerly possessed a large priory, as is indicated by the designation of the home of the Delmé-Radcliffes; as well as by the existence in the town itself of certain almshouses known as the "Biggin." The latter were

Ruins of Sopwell Nunnery, St Albans

purchased by a private gentleman in 1545, being at that time part of the disestablished "Priory of Bygyng in the town of Hychen." The Biggin, which was once inhabited by Gilbertine nuns, has a beautiful wooden corridor. "The Priory" as the title of a house in the main street of Redbourn, and "The Cell" as that of a mansion further down the road, at Markyate, are but two among many other traces of monastic institutions in the county.

Of King's Langley Priory, which is known to have been in existence in 1400, a considerable portion still

The Priory, Hitchin

exists. Ashridge House now occupies the site of a large monastery and college, of which there are many remains. A brief reference may here be made to St Albans' clock-tower, which was erected between 1403 and 1412, and from which the curfew was rung till so late as 1861,

Courtyard in the Biggin Almshouses, Hitchin

while a bell was also rung early in the morning to awaken work-people.

17. Architecture. (*b*) Military—Castles.

Like most other counties in the south of England, Hertfordshire possesses the remains of several Norman

The Priory, King's Langley

castles, most of which appear to date back no further than the Conquest, while others, like Berkhampstead (where, as we have seen in a previous section, Mercian kings held their courts), have been supposed to be constructed on the site of earlier buildings of a similar nature.

The total number of castles built by the Normans to

overawe their new English subjects is stated to have been about 1100. These, as may naturally be surmised, varied considerably in size, some being royal castles, constructed for the defence of the country generally and ruled by a constable or guardian, while others belonged to individual Norman noblemen for the defence of their own estates, and were for the most part the terror of the surrounding districts.

A Norman castle of the highest type occupied a large area, the lofty and massive outer wall enclosing a space of several acres, and being surmounted with towers and protected by bastions, while it was also surrounded by a moat or ditch. Within the enclosure thus formed were three main divisions, the first of which was the outer bailey, or courtyard, entered by a towered gateway furnished with a portcullis (that is to say, a gate which could be dropped down from, and drawn up into, the tower by means of a system of chains and pulleys) and a drawbridge. The stables and other buildings were contained in this court. Next came the inner bailey, or quadrangle, likewise entered through a towered and fortified gateway, and containing the chapel, the barracks, and the keep. Lastly, we have the keep or donjon itself, which always contained a well, and constituted the final portion which was defended during a protracted siege when the garrison was hard pressed. In choosing the site for such a military castle, either a more or less isolated and steep hill or rock might be selected, or a situation in marshy low-lands, where access might be rendered difficult or impossible by damming back the waters.

The Norman castle at Berkhampstead, which stood close to the present railway due east of the station, and portions of the ruins of which may be seen from the train, was built by Robert Earl of Morton, brother of William the Conqueror.   According to a recent writer, the earthworks of this castle represent the original fortress founded by the Conqueror, and the appellation of a "burh" to the structure is consequently erroneous.   A Saxon "burh" or "burg" was a fortified town, whereas the moated mound of Berkhampstead, like those at Hertford, Bishop's Stortford, Anstey, Bennington, and Pirton, are essentially Norman castles of the type known as "mottes," or, from their shape, as "mount and bailey castles."   It is a common idea that Berkhampstead was originally a stone castle, but the earthworks now remaining really represent the fortress itself.   In the reign of Henry II the custody of Berkhampstead Castle was entrusted to Thomas à Becket, who replaced the old wooden defences (such as stockades, palisades, and towers) originally crowning the banks, by walls of flint rubble, remains of which still partly surround the enclosure.

Hertford Castle, the site of which forms the residence of His Majesty's judges during the assizes, was built by Edward the Elder about the year 905; and after the conquest William I placed both castle and town in the custody of Peter de Valoignes.   Other ancient castles and baileys in the county include the following, viz :—

Anstey Castle, situated about a mile from the eastern border of Hertfordshire, on the watershed between the Stort and the Quin.   Bennington Castle, built on high ground about a mile from the river Beane and some two miles from Walkern Bury; the ruins include the remains of a small, square keep, as well as of a bailey. Wayte-more Castle, Bishop's Stortford, belonging to the Bishop of London, is an excellent example of the type of fortress which owes the main part of its strength to being situated in a practically impassable morass; that is to say, when the latter was kept well flooded. The flint rubble walls of the keep are fully a dozen feet in thickness.

Smaller baileys existed at High Down, near Pirton; at Periwinkle Hill, on nearly level ground, midway between Reed and Barkway; and also at Walkern, on the Beane, about a mile and a half distant from the village. In the last of these it is noteworthy that the church is situated close to the castle, although, unlike the one at Anstey, it does not appear to have been included in an outer ward.

The class of defensive works known as homestead moats—that is to say, simple enclosures formed into islands by means of moats containing water—do not, perhaps, strictly speaking, come under the title of either architectural or military structures. Still this seems the most convenient place in which to mention them. The northern and eastern districts of Hertfordshire are remark-able for the enormous number of these homestead moats; these districts being equalled in this respect only by Essex and Suffolk. On the western side of the county they are comparatively uncommon.

"These enclosures," observes a writer in the *Victoria History* of the county, "vary greatly in size, shape, and position, and it is obvious that they do not all belong to one period, for in all ages to surround a piece of land with a ditch has been one of the most elementary forms of defence. There are, however, as with the larger earthworks, certain typical forms....It should be noted that the typical feature of a homestead moat is that the earth, dug to form the deep surrounding ditch, was thrown on to the inclosure and spread, thus raising the island slightly above the surrounding level. The construction of moats, except for ornamental purposes, having ceased when the state of the country no longer necessitated such protective measures against men or wild beasts, they often fell into decay, or were partially filled up, and their vestiges converted into ponds, while many may have been obliterated as interfering with agriculture, but there still remain a large number."

To reproduce the list of these would occupy far too much space; and it must suffice to mention that examples of homestead moats may be seen in the parishes of Ashwell, Braughing, Pirton, and Sawbridgeworth.

A step in advance of the homestead moat was formed by earthworks made on the same plan as the latter, but provided with a rampart and a "fosse," or ditch, and in some cases also with outer defences. Of this type of earthwork three examples are definitely known to occur in the county, namely one at Bygrave, a second at Whomerley Wood to the south-east of Stevenage, and a third at Well Wood, Watton.

Yet another kind of defensive earthwork is to be found in the shape of walls, ramparts, or ditches surrounding the sites of ancient villages. Of this we have a local example in Kingsbury Castle, an old fortified village lying to the south-west of the city of St Albans, and covering an area of about 27½ acres. The village stood upon a hill, of which the summit has been planed off and the material employed to form steep banks or ramparts, one of which was partially thrown down to form the present Verulam Road, while another portion persists in the shape of a steep fall in the gardens or yards at the back of the houses on the north side of Fishpool Street. The main structure of the castle was demolished during the tenth century, and the remnant about the year 1152. It may be added that the clay-pits on the north side of Kingsbury Castle are the reputed source of the material of the Roman bricks of which Verulam is built.

The gigantic earthworks of the type of Beech Bottom and Grimm's Dyke have been mentioned in an earlier section.

18.   Architecture.   (c)   Domestic — Famous Seats, Manor Houses, Cottages.

With the advent of less troublous times at the close of the Wars of the Roses a marked change is noticeable in the plan and architecture of the residences of the great noblemen and country gentlemen. The need for castles or fortified houses ceased to exist; and attention was consequently directed to comfort rather than strength in the construction of country mansions. Fortunately a number of these fine old Tudor residences have survived in different parts of the country; but many have been replaced by other later structures built on the old foundations.

These Tudor mansions usually took the form of a large house built round a quadrangle, the hall occupying the middle portion of the building, with flanking wings on both sides. The building material depended upon the locality and on the taste and means of the owner; but in this county brick was extensively employed by the Tudor, and still more so by the Stuart builders.

In lordly country seats, as well as in mansions of a less pretentious type, dating from the Tudor period downward, Hertfordshire, owing doubtless to its well-wooded and picturesque scenery, its good soil, bracing climate, and proximity to the metropolis, is especially rich, and in this respect presents a marked contrast to the neighbouring county of Essex. The majority of these houses, however, have been either completely rebuilt or more or less extensively altered at later epochs.

Among the few of these noble residences that can be mentioned here, Hatfield House, which was built between the years 1605 and 1611 by the first Earl of Salisbury, presents a magnificent specimen of early Jacobean architecture in brick and stone, mellowed by time to exquisitely soft tints. The original palace, where Edward VI lived, and where Elizabeth was kept in captivity, now forms the stables. The mention of the virgin queen naturally leads on to Ashridge, near Berkhampstead, formerly the seat of the Dukes of Bridgewater, where Elizabeth also spent a considerable time in her early days. Although the present building, which stands partly in Hertfordshire and partly in Buckinghamshire, is mostly modern Gothic, the fine vaulted cellar is a remnant of the old monastery and college which formerly occupied the site. Knebworth, near Stevenage, the home of the Earls of Lytton, although

Hatfield House, South Front

now a comparatively modern Gothic building, was originally a Tudor mansion, dating from the reign of Henry VII, the present house occupying the position of one of the four wings of the original building. Tittenhanger, between St Albans and Colney, occupies the site of a royal residence dating from the fourteenth and early part of the fifteenth century; the present mansion, notable for its grand oak staircase, is stated to have been built in 1654, although the style of the brick-work suggests the early part of the eighteenth century. Little Hadham Hall, at the village of that name, is a splendid example of Elizabethan architecture in red brick.

Knebworth

A very interesting mansion is Salisbury House, Shenley, built some time before 1669; much of the original brick building still remaining as an excellent example of Stuart architecture. The house is surrounded by a broad moat, and is approached by a bridge. Mackery End, near Wheathampstead, contains some fine examples of sixteenth century architecture; and Rothamsted, near Harpenden, is in the main a seventeenth century brick mansion, dating from between 1630 and 1650, although it has older portions, and the hall belonged to a house constructed of timber on a flint base.

Of fine old houses now forming farm-homesteads there are many examples on the western side of the county.

Water End Farm near Wheathampstead
(*Elizabethan Manor-House*)

Among these is Turner's Hall, to the north-west of Harpenden, now considerably modernised.

Another very interesting building of this type is Water End Farm, in the parish of Sandridge, situated on the banks of the Lea about two miles from Wheathampstead, and stated to have been built about 1610. It

is constructed of brick and has the straight-gabled, mul-
lioned style characteristic of the later part of the reign of
Elizabeth and the commencement of that of James I.
Like many other houses of Elizabethan times it is
constructed in the form of the letter E, and its three
stacks of brick chimneys, with octagonal shafts and

Christ's Hospital School, Hertford

moulded brick caps and bases, are especially characteristic
of the style of this period.

Of later date are the Marlborough Buildings, or
Almshouses, St Albans, erected by Sarah, Duchess of
Marlborough in 1736, and affording a fine example of
the brick architecture of that period in an excellent state
of preservation. Here also may be mentioned the Blue-

Coat School at Hertford, and the Boys' Grammar School at Hitchin.

Forty years ago the county abounded in picturesque brick-and-timber cottages, roofed with either tiles or thatch; but these are disappearing yearly under the hand of the speculative builder, to be replaced by hideous box-

The Grammar School, Hitchin

like buildings of brick and slate. Some, however, still survive, either in the towns or the smaller hamlets, such as picturesque Amswell, near Wheathampstead, which may be cited as an ideal example of one of the smaller Hertfordshire villages.

As has been well remarked in another volume of the present series, the great difference between these ancient cottages and houses and the great majority of their modern successors is that while the former harmonise with their surroundings, reflect not a little of the spirit of the builder, and improve, like good wine, with age, the latter are altogether out of keeping, and are likely to become, if possible, still more offensive and objectionable with the advance of time.

An Old Malting House, Baldock

While most of the old Hertfordshire cottages were of brick and timber, others were built of flint with brick facings, or more rarely of rounded pebbles from the Woolwich and Reading beds, or with brick courses and window-mullions; some were of feather-edge boarding, and others again of rubble and plaster.

In Hemel Hempstead High Street is a building, now converted into cottages, which contains above the fire-places on the ground and first floors the Tudor rose and

Chequer's Yard, Watford

fleur-de-lys in plaster-work; while the back of a neigh-bouring building probably dates from the time of Henry VIII. Excellent examples of the old brick-

and-timber cottages are to be seen in the village of Northchurch, and also at Aldbury, east of Tring, where the old parish stocks are likewise preserved. Most of these Aldbury cottages are tiled, although a few are

The "Fighting Cocks," St Albans
(Ancient Inn near the Ford across the Ver)

covered with thatch, a style of roofing much less common in that district than in many parts of the county. Watford has still a number of old cottages, notably in

Farthing Lane and Chequer's Yard, and in St Albans, especially in the market-place and French Row, there are several dating from the sixteenth and seventeenth centuries. There is also a very remarkable old hexagonal wooden house near the ford across the Ver at St Michael's silk-mills, said to be one of the oldest licensed houses in England.

Waltham Cross

Did space permit, reference might be made to old houses in Hertford, Berkhampstead, and other towns and villages, but the facts mentioned are sufficient to indicate the interest of the county to antiquarians in the matter of ancient buildings, and before concluding this section we must not omit to mention what is certainly not the least

interesting of all Hertfordshire antiquities—the Cross at
Waltham which Edward erected to his beloved Queen
Eleanor ; the last before arriving in London of the fifteen
commemorating the resting-place of her body on its
journey from Grantham to Westminster Abbey in 1290.

19. Communications—Past and Present. Roads, Railways, Canals.

Lying as it does on the direct route from the metropolis
to the north and north-west of England, and containing
in its western portion the formerly important city of
Verulam, Hertfordshire, as might be expected, is traversed
by several trunk roads leading in those directions, two
of which date from Roman times.   What these lines of
communication were in pre-Roman days we have no
means of knowing, although it is probable that they were
little more than rude tracks through the great forest, or
"weald," which in those days extended some forty miles
to the north of London, and afforded shelter to the great
wild ox, red deer, wild boars, bears, and wolves.   Road-
making was a special attribute of the ancient Romans ;
and after they had constructed highways from their early
stations in Kent, they probably set to work on those in the
counties to the northward of London.   So well made and
so straight were these ancient Roman roads that many of
them (with in some cases a certain amount of local deviation)
have remained the main highways of the country down to
the present day.   Immediately before the introduction of

railways, when the coaching traffic was brought to its
highest pitch of development, these main trunk roads—
thanks to the invention of Macadam—were maintained
in superb condition, though with the extension of the
railway system some of them were allowed to deteriorate.

There are three great Roman roads traversing
Hertfordshire—Watling Street, the Icknield Way, and

The Ermine Street at Hertford Heath

Ermine Street. Watling Street starts from Dover, and
after passing through London, enters the county to the
south of Elstree, whence it is continued through Colney
Street, Park Street and St Stephens to St Albans, and
thence on through Redbourn and Markyate Street, and
so to Dunstable whence it eventually reached Chester

The Icknield Way, showing a Ford between Ickleford
and Wilbury Hill

and Holyhead. Although frequently miscalled the North Road, the modern representative of Watling Street is known as the Chester and Holyhead road. Originally the Roman road in the neighbourhood of St Albans ran altogether to the west of the Ver, from Gorham Block to the Pondyards; this section of the modern road, which crosses the river to enter the city, having been constructed during the years 1826—1834.

The Icknield Way (taking its name apparently from the British tribe of the Iceni) may be the oldest of the three great tracks, and originally of pre-Roman age, as, like the Pilgrims' Way in Kent, it mainly follows the line of the chalk downs. It may be called a cross-country road from the west of England, cutting the Watling Street at Dunstable, and thence extending in a north-easterly direction across Hertfordshire through Little Offley, Ickleford, and Baldock, and thence by way of Royston, where it crosses the Ermine Street, to Newmarket and Yarmouth.

The Ermine Street, the third great Roman road, takes, on the other hand, a northern direction, passing through Cheshunt, Wormley, Broxbourne, and Wadesmill, and so by way of Buntingford to Royston. There is however some difference of opinion about its course.

Of modern roads, the Chester and Holyhead road has been already mentioned as following in the main the line of the Old Watling Street. Of equal importance is the Great North Road to York, passing through Barnet, Hatfield, Welwyn, Codicote, Stevenage, and Hitchin. Of other highways it must suffice to mention that the

High Street, Stevenage

Bedford road branches off from the Chester and Holyhead at St Albans to run through Harpenden, and so on to Luton, in Bedfordshire; while the main road from London to Cambridge and Norwich takes the line of the Lea valley on the eastern side of the county, which it leaves a short distance to the northward of Bishop's Stortford.

View on the 'Great North Road,' Codicote Village

In coaching days St Albans was a far more bustling and busy town than it is at the present day; a very large number of coaches passing daily through the city each way, the majority running on the Holyhead and Chester road, but a certain number taking the Bedford line. The two chief coaching and posting inns were the Peahen and the White Hart, both of which are still in existence.

The speed and smartness with which the mail-coaches were run in the days immediately preceding their abolition was little short of marvellous.

The first half of the nineteenth century saw the introduction of railways, which gradually but surely killed the old coaching traffic. One of the first lines to be

Watford

opened was the London and North-Western—in 1838—which traverses the south-western side of the county, passing through Watford, Boxmoor, Berkhampstead, and the outskirts of Tring. In 1853 a branch line was opened from Watford to St Albans, and another to Rickmansworth in 1862. With a short break between Boxmoor and Hemel Hempstead, the North-Western system is connected with the Midland by means of a branch line from the last-named town to Harpenden. The main line of the Midland, which traverses the western half of the county by way of Elstree, St Albans, and Harpenden, was opened in 1868. Both St Albans and Harpenden have branches of the Great Northern Railway to Hatfield, which is on the main line; the latter continuing through Welwyn, Stevenage, and Hitchin. By means of a branch line of the Great Northern from Hatfield to Hertford, we reach the Great Eastern, the fourth great railway in the county, the main line of which runs through Broxbourne, Sawbridgeworth, and Bishop's Stortford, but connected also with Ware and Hertford, and having a branch from Stortford to Buntingford.

With such a multiplicity of lines, it might well be imagined that railway communication between nearly all parts of the county would be well-nigh perfect. As a matter of fact, this is by no means the case; and the journey by rail from the western to the eastern side, owing to changes and delays, is so slow and tedious, that it is frequently found convenient to hold important Hertfordshire meetings, like those of the County Council, in London.

As regards water-communication, the western side of the county is served by the Grand Junction Canal, which, after leaving Leighton Buzzard, enters the county near Tring, and thence runs by way of Berkhampstead, Boxmoor, Hemel Hempstead, Runton Bridge, Watford and Rickmansworth in a south-easterly and southerly direction to London. A considerable amount of barge-traffic is still carried on on this canal, although nothing approaching that in pre-railway times. In those days the whole of the coal-supply for north-western Herts came by canal to Boxmoor, whence it had to be carted for long distances—some 14 miles to Harpenden, for instance. As there are at least two very steep hills—which become impossible for heavily laden teams when the roads are slippery—between Boxmoor and Harpenden, the inhabitants of the latter

The Grand Junction Canal near Hemel Hempstead

picturesque village were apt to run short of firing at Christmas.

On the other side of the county the Lea is navigable for barges as far up as Ware and Hertford; and here too a considerable amount of heavy traffic is still carried on by water.

In this place mention may conveniently be made of

Haileybury College

the New River, running from the valley of the Lea near the Rye House, at a gradually increasing distance from that river, to the metropolis. The New River, or Middleton's Waters, as it used also to be called, was constructed in the reign of James I, at first almost entirely by Sir Hugh Middleton, but later on by a company with a special charter, for the purpose of supplying north London with drinking-water. The chief sources of the New River are the springs at Chadwell and Amwell. At the present time an original £100 share in the New River Company is worth an almost fabulous price.

20. Administration and Divisions— Ancient and Modern.

The present administration and administrative divisions of Hertfordshire, like those of other English counties, have been gradually evolved and developed from those of our Saxon forefathers; each alteration in the form of local government and of local administrative boundaries being based on the previously existing system. By the Saxons each county was divided into a number of main divisions known as hundreds, or wapentakes, each governed by a hundreder, or centenary (the equivalent of the Old German *Zentgrafen*), and each having a name of its own. Hertfordshire is now divided into eight hundreds, the names of which, commencing on the western side of the county, are as follows: Dacorum (including Tring), Cassio (with the important towns of St Albans, Watford, and Rickmansworth), Hertford, Braughing, Broadwater (occupying nearly the centre), Hitchin and Pirton (on the north-west corner), Odsey (in the extreme north), and Edwinstree (on the north-east). Originally they were more numerous, Cassio, for instance, being much smaller than at present, while the Hitchin division was reckoned only as a half-hundred. The origin of the names of most of the hundreds are self-apparent; but that of Cassio (originally Kayso) appears to be unknown, while that of Dacorum has some connection with the Danes, perhaps referring to a Danish settlement.

One of the most remarkable facts connected with the hundreds of Hertfordshire is that three of them do not lie within what farmers call a ring-fence. Dacorum, for instance, has two outlying areas in the south-eastern corner of Cassio, and a third wedged in between Cassio on the west, Broadwater on the north, and an outlying portion of Cassio on the east. Broadwater, again, has a small outlier on the Middlesex border of the south-eastern "peninsula" of Cassio; while Cassio itself, inclusive of the one already mentioned, has no less than eight of these curious outliers, one situated in the extreme north in the hundred of Odsey.

Each hundred originally had its own court, or "hundred-mote," which met monthly; and it was divided, as at present, into townships, or parishes. The parish, in turn, had its own council, or *gemot*, where every freeman had a right to appear. This assembly or council made its own local by-laws, to enforce which it had a reeve, a bailiff, and a tithingman, with the powers of a constable.

The reeve was chairman of the township *gemot*, and could summon that assembly at pleasure.

Passing on to more modern times, we find Hertfordshire occupying a peculiar position in regard to local government and administration in that it possessed a kind of *imperium in imperio* in the shape of what was known as the Liberty of St Alban; in other words, a large area on the western side of the county originally under the jurisdiction of the abbots of St Albans, who had the power of inflicting the death-penalty. Originally there was a separate Commission of the Peace for the Liberty, so that a Justice for the County had no jurisdiction in the former unless he had been specially inducted. This arrangement was found, however, to be inconvenient, and the Liberty, as such, was abolished, although it was taken as a basis for the splitting of the county into a western and an eastern division for judicial purposes.

The chief officers of the county are the Lord Lieutenant and the High Sheriff; the former (who in Hertfordshire is always a nobleman) being the direct local representative of the sovereign, and having the appointment of magistrates and the officers of the territorial forces, while the latter (who is a commoner) is the head of the executive department in the administration of justice. The Lord Lieutenant holds office for life, or during the sovereign's pleasure, but the Sheriff is appointed annually by the Crown. Deputy Lieutenants are supposed to act, in case of need, for the Lord Lieutenant.

Formerly the greater part of the business of the county was conducted by the Justices of the Peace, or

Magistrates, at Quarter Sessions, but most of this is now transferred to the County Council, which, as previously stated, often meets in London. This County Council, which was first established in 1888, is composed of Aldermen and Councillors; the latter of whom are elected, while the former are what is called "co-opted," that is to say, selected by the Council itself, either from its own body, or from the general public. The duties of the County Council include the maintenance of high roads and bridges; the appointment and control, in conjunction with the magistrates, of the police; the management of reformatories and lunatic asylums; and, in a word, the general carrying out of the laws enacted by Parliament.

According to a scheme elaborated in an Act of Parliament passed in 1894, the more important minor local bodies are denominated District Councils, and those whose function is less Parish Councils; the former having control of the more populous towns and villages, other than cities and boroughs, and the latter those with fewer inhabitants. For this purpose many parishes are divided into a more populous Urban and a less populous Rural District. Certain towns in the county rank, however, as cities, or boroughs, and have larger powers and different forms of government; being ruled by a Mayor and Corporation, and having magistrates and a police force distinct from those of the county. Among these privileged towns, St Albans ranks as a city, while Hertford and Hemel Hempstead are boroughs. Hemel Hempstead is a very ancient borough, and has, in addition to its Mayor, an official known as the High Bailiff.

The county is likewise divided into a number of Poor Law Unions, each with a Board of Guardians, whose duty it is to manage the workhouses, and appoint officers to carry out the work of relieving the poor and those incapacitated by age or other cause from earning their own living.

The Shire Hall, Hertford

As regards the administration of justice, Assizes are held by His Majesty's Judges three or four times a year at the Shire Hall, Hertford, for the whole county; the Grand Jury on such occasions being composed entirely, or mainly, of magistrates.   Quarter Sessions, on the other hand, are held four times a year at Hertford for the eastern, and at the Court House, St Albans, for the

western division of the county ; these courts being constituted by the magistrates for the county and the mayors
of the boroughs and city. Petty Sessions are held weekly,
fortnightly, or monthly at a number of the towns and larger
villages. In most cases the county magistrates in the
immediate neighbourhood preside at these sessions ; but
the city of St Albans and the two boroughs have magistrates of their own, who also hold petty sessions for trying
cases which occur within the area of their jurisdiction.

St Albans is the centre of an episcopal diocese, which
includes most of that portion of London situated within
the county of Essex. Arrangements are, however, now
in progress for relieving the Bishop of St Albans of the
care of that part of the diocese commonly known as
" London Over the Border."

The diocese, so far as Hertfordshire is concerned, is
divided into archdeaconries, rural deaneries, and parishes.
The latter are very numerous, although somewhat less
so than the civil parishes, for the purposes of which, as
already mentioned, the ecclesiastical parishes are frequently
split into an urban and a rural section. There are 170
ecclesiastical parishes situated wholly or partly within the
old county, of which 164 are included in the diocese
of St Albans ; while three belong to Ely, two to Oxford,
and part of one (Northwood) to London.

The larger towns, the city, and the two boroughs
have Education Committees of their own ; but for the
rest of the county a Committee of this nature is appointed
by the County Council.

Hertfordshire has four parliamentary divisions, namely,

Hertford, Hitchin, St Albans, and Watford, each of which returns one member to the House of Commons. The county is thus represented only by four members, as against fifteen for Kent.

21. The Roll of Honour of the County.

Hertfordshire cannot hope to rival such counties as Norfolk or Kent in its roll of distinguished names, but it can show a fairly long list of persons connected with the county who have been famous.

Since reference has already been made in several of the foregoing sections to the visits of English sovereigns to the county, or to their residence within its borders, very brief mention of the connection between royalty and the county will suffice in this place. Neither here nor elsewhere in these pages is any attempt made to give a complete list of such visits.

The names of Boadicea, queen of the Iceni, and of Offa, king of Mercia, who had his palace at Offley, dying there in 796, will always be specially connected with Hertfordshire. In a somewhat less degree the same may be said of William the Conqueror, to whom, as already mentioned, the crown of this realm was offered at Berkhampstead. Edward II and Edward III frequently resided at Langley Palace, where Edmund de Langley, the founder of the White Rose faction, was born in 1341; and the same residence was also used by Richard II. Henry I and his consort Matilda were present at the dedication of St Albans' Abbey on its completion by Abbot Paul; and Henry VI was at the first battle of St Albans, where he was wounded. Henry VIII, as mentioned on page 82, was still more intimately connected with Hertfordshire, and the manor of Hitchin was conferred by him in turn on Anne Boleyn and her successors. Reference has already been made to the residence of Queen Mary, in her youth, at Ashridge, and of Queen Elizabeth (before her ascent to the throne) both there and at Hatfield; while, as sovereign, Elizabeth also visited St Albans on two or three occasions as the guest of Sir Nicholas Bacon at Gorhambury, and also went to other great houses in the county. James I, as mentioned on the same page, spent much time at Royston, and died at Theobalds. The Rye House plot, so called from the meeting-place of the conspirators at Broxbourne, as stated in an earlier section, was devised for the purpose of assassinating Charles II while on his way through the county.

In connection with personages of royal blood, mention may be made of Humphry, Duke of Gloucester, whose name is so intimately associated with St Albans' Abbey, to the monastery of which he was admitted a member in 1423; and also of Sarah, Duchess of Marlborough, who was born at Sandridge in the eighteenth century, and built and endowed the almshouses bearing her name in St Albans.

Among great statesmen connected with the county a prominent place must be assigned to Queen Elizabeth's councillors and favourites, Lord Burleigh and the Earl of

Essex. To her reign likewise belongs Sir Nicholas Bacon, Keeper of the Great Seal, and owner of Gorhambury, where he died in 1578. Nearly a century later (1652), Gorhambury came into the possession of Sir Harbottle

The Salisbury Statue, Hatfield

Grimston, well known as Speaker of the House of Commons.

Passing on to the Victorian age, we have two great statesmen, namely, Lord Melbourne and Lord Palmerston, both of whom lived at Brocket, where the former died;

and, subsequently, the late Marquis of Salisbury, owner of stately Hatfield. The late Viscount Peel, sometime Speaker of the House of Commons, was also a Hertfordshire man, with his residence at Kimpton Hoo. Cecil Rhodes,

Cecil Rhodes's Birth-place, Bishop's Stortford

the South African premier and "Empire-builder," likewise claims a place in the roll of honour of the county, having been born at Bishop's Stortford rectory, and Commodore Anson, the great circumnavigator, though not a native, lived at Moor Park, where he died in 1762.

Dame Juliana Berners, imaginary prioress of Sopwell

nunnery, who was supposed to have written the immortal *Treatyse on Fysshynge with an Angle*, the first work on angling ever published in England, has been shown to be a myth. Among names famous in literature and science the greatest connected with the county is perhaps that of the great philosopher Sir Francis Bacon, afterwards Lord Verulam and Viscount St Albans, who,

Ruins of Verulam House, the Residence
of Francis, Viscount St Albans

during his father's residence at Gorhambury, lived in Verulam House, at the Pondyards. On the death of his father Sir Nicholas Bacon he succeeded to Gorhambury. By a curious error he is frequently called Lord Bacon, although no such title was ever in existence. John Bunyan claims a place among Hertfordshire literary worthies as he was connected with a chapel at Hitchin.

Two of the greatest literary names connected with the county are those of William Cowper the poet, and Charles Lamb, author of the *Essays of Elia*, unsurpassed as a master of delicately humorous prose, whether as essayist or letter-writer. The former was born at

Francis Bacon, Viscount St Albans

Berkhampstead rectory in the year 1731; but Lamb was chiefly a visitor to the county, though, as he tells us in the *Essays*, he was once a Hertfordshire landowner, and his cottage at West Hill Green, about 2½ miles from

Puckeridge, still exists. Mackery End Farm was the residence of the Brutons, who were his relatives, and it was to their house that his visits were made ; so that the neighbourhood is essentially Lamb's country. It is a question whether the Lyttons or the beauties of Knebworth, their home, are the more famous. The great novelist, author of *The Last Days of Pompeii*, *The Caxtons*, and innumerable other tales, as well as such successful plays as *The Lady of Lyons* and *Money*, was best known to readers' in the middle of the last century as Sir Edward Bulwer-Lytton, though he began life as Mr Bulwer and died Lord Lytton. His son, poet, Ambassador, and Viceroy, who wrote under the name of "Owen Meredith," was scarcely less distinguished, and received an Earldom in 1880. Here, too, mention may be made of Mrs Thrale, the friend of Dr Johnson, who was often at Offley Place, where her husband, whose family was long connected with St Albans, was born. Offley Place was at this time a fine old Elizabethan mansion, although it has since been rebuilt. Gadebridge Park, Hemel Hempstead, was the residence of the great surgeon Sir Astley Paston Cooper. But a greater distinction attaches to the name of Rothamsted, near Harpenden, as being the residence of the late Sir John Bennet Lawes, Bart., who, with his scientific colleague Sir Henry Gilbert, conducted the experiments which made their names famous throughout the agricultural world. Sir John Lawes first obtained the idea of using fossilised phosphates for manure from Professor Henslow, the great Cambridge botanist (himself sometime a resident at Hall Place, St Albans), who sent him specimens obtained from the Essex "Crag," with a suggestion that they might be used as a source of phosphoric acid. Yarrell, the naturalist, lies buried in

Charles Lamb

Bayford churchyard, with many members of his family. Last in the scientific and literary list, we have the name of Sir John Evans, the great antiquarian and numismatist of Nash Mills, Hemel Hempstead, who died so recently as 1908. To Evans, in conjunction with the late Sir Joseph Prestwich, is mainly due the credit of definitely establishing the fact that the so-called flint "celts" are really the work of prehistoric man. His most important and best work is *Ancient Stone Implements*.

William Cowper

Among great ecclesiastics mention must be made of Nicholas Breakspear, born near Abbot's Langley towards the close of the eleventh century, who subsequently became Pope as Adrian IV; being the only Englishman who has occupied the papal chair. Reference may also be made to Cardinal Wolsey, who spent a considerable portion of his time at Delamere House. Nor must we omit Young, the author of the *Night Thoughts* and Rector of Welwyn, or that great maker of hymns, Dr Watts, who as the 36-year guest of Sir Thomas Abney resided at Theobalds, where he died. Among distinguished lawyers, the most prominent name is that of Lord Grimthorpe (formerly Sir Edmund Beckett), who was, however, connected with the county, not in his professional capacity, but as the restorer of St Albans' Abbey and other churches in the neighbourhood. Much criticism has been expended on Lord Grimthorpe's modes of "restoration," which were certainly of a drastic character. It must, however, be remembered that when he undertook the restoration of St Albans' Abbey it was in a dangerous condition, and sufficient money was not forthcoming to make it secure. The result is that the abbey, although in many ways unlike its former self, will stand for centuries. Lord Grimthorpe, who was a Yorkshireman, built himself a residence at Batchwood, near St Albans.

As another well-known lawyer and also a judge, mention may be made of Lord Brampton (Sir Henry Hawkins), who came of a family long connected with Hitchin, at which town he was born.

Sir Henry Chauncy, the antiquary and historian of the county, to whom reference has so often been made in this book, lived, and died in 1700, at Yardley; and Balfe, the composer, made his home at Rowney Abbey, close by, till his death in 1870.

22. THE CHIEF TOWNS AND VILLAGES OF HERTFORDSHIRE.

(The figures in brackets after each name give the population in 1901, and those at the end of the sections give the references to the text.)

Abbot's Langley (3342), a village situated on the Gade, with a station on the North-Western Railway; it was bestowed in the time of Edward the Confessor upon the then abbot of St Albans, whence its name. Hunton Mill, on the Gade, was granted to Sir Richard Lee in 1544, and both this and Nash Mills were farmed from the abbot of St Albans between 1349 and 1396. The present church, dedicated to St Lawrence, cannot be traced farther back than the close of the twelfth century. A west tower was added about 1200. (pp. 72, 90, 105, 149.)

Aldenham (2437), a village and manor lying to the north-east of Watford. It has a grammar school; and near by is Aldenham Abbey, the seat of Lord Aldenham. In 1898 two Roman kilns were discovered in the parish. The church, which has been restored, contains one small twelfth-century window; no trace of the chancel remains.

The Amwells—Great Amwell (1421), and **Little Amwell** (930)—small villages not far from the Rye House. Amwell is associated with the name of the Quaker poet, John Scott, who lived there for some time after 1740. Near by is Haileybury, formerly the training college for the officials of the East India Company, but now a public school. (p. 135.)

Ashridge, a domain in Little Gaddesden parish, situated on the Buckinghamshire border of the county, and celebrated for its splendid beech woods. It was formerly the property of the Dukes of Bridgewater, being acquired by the Egertons in 1604, but it is now owned by Earl Brownlow. A building, formerly the porter's lodge, includes some remains of an old monastic college. The present house, which stands partly in Buckinghamshire, was built by the eighth Earl of Bridgewater. (pp. 115, 142.)

Ashwell (1281), a village on the Cambridgeshire border of the county, with a station some distance away on the Royston and Cambridge branch of the Great Northern Railway. Ashwell, which was formerly a town, had a fair and a market in the time of William the Conqueror. It was severely visited by the plague. Its church-tower is the only one in the county built wholly of stone. (pp. 92, 113.)   ə

Baldock (2057) is a market-town on the Icknield Way, to the north-west of Hitchin, with a station on the above-mentioned branch of the Great Northern Railway. It dates from Norman times, when it was known as Baudok. During the Crusades, Baldock, like St Albans, Berkhampstead, and Hoddesdon, had a lazar-house for lepers, who were at that time numerous all over England. The list of Rectors is complete from the days of the Knights Hospitallers in 1317. The church contains much Decorated and Perpendicular work. (pp. 90, 94, 128.)

Barkway (661), originally Berkway, is a small town and manor situated a few miles to the north-east of Buntingford. (p. 96.)

Barnet, or **Chipping Barnet,** originally Chipping Bernet (7876), a large and important market-town near the Middlesex border of the county, with a station (High Barnet) on a branch of the Great Northern Railway. Near by are New Barnet and East Barnet, with a station on the main line, and having a population of 10,024. The name Barnet is a corruption of the Saxon

Bergnet, signifying a little hill; the site of the town then forming a small rising in the midst of the great forest; the prefix Chipping =market is a word of Scandinavian origin, represented in the Swedish Jonköping and the Danish Kjøbenhavn = (Copenhagen). Barnet has a castle and was the scene of a battle in 1471, when the Yorkists defeated the Lancastrians, killing their leader, Warwick the king-maker. The market was famous for its cattle; and in addition to this there is an annual horse-fair, which formerly attracted dealers from all parts of the country. (pp. 9, 53, 54, 82, 99, 128.)

East Barnet (2867), known as La Barnette in the thirteenth century, and Low Barnet in the fifteenth century, is situated on the stream known as Pymmes' Brook, on the western side of the valley of which stands the almost deserted old parish church.

Bayford (330), a village nearly midway between Hatfield and Hoddesdon. In the churchyard is buried William Yarrell the naturalist. Bayfordbury is celebrated for its collection of portraits of members of the Kitcat Club painted by Sir Godfrey Kneller. (p. 147.)

Bengeo (3063), formerly Bengehoo, a village in the valley of the Beane one mile north of Hertford. The old church, now little used, is one of the oldest in the county, dating apparently from the early Norman period. Together with Great Wymondley church, it is peculiar, so far as Hertfordshire is concerned, in having an apsidal chancel. In place of a tower, it has a wooden bell-cote. Panshanger, formerly the property of the late Lord Cowper, is near by.

Bennington (522), a market-town and manor, situated on the Beane, from which it takes its name; it was an important place in the ninth century, when it was the residence of the kings of Mercia. The church is fourteenth century. (pp. 53, 54, 56, 78, 111.)

Berkhampstead, or **Berkhampstead Magna** (5140), an important market-town on the London and North-Western Railway and Grand Junction Canal, and one of the oldest in the county, the castle dating from Norman times, and being possibly on the site of an earlier Saxon edifice. It was here that the crown of England was offered to William the Conqueror. The manor and castle were granted first to Piers Gaveston and subsequently to Edward the Black Prince, but were afterwards annexed to the Duchy of Cornwall. Berkhampstead, which is now a petty-sessions town, and has an ancient grammar school, formerly returned burgesses to Parliament. Its almshouses were built in 1684. Cowper was born here. Berkhampstead Parva and Berkhampstead St Mary's—the latter now generally known as Northchurch—are villages in the neighbourhood. The church dates from the thirteenth century; it contains the beautiful Torrington tomb. (pp. 2, 7, 63, 80, 81, 82, 91, 92, 109, 111, 131, 132, 141.)

Bishop's Stortford (7143) is the most important town on the Essex border of the county, and has a station on the main line of the Great Eastern Railway, and a bridge over the Stort. The town, which has a market, possessed a considerable trade in Saxon times, and was the property of the Bishop of London, and to protect this, and for the purpose of consolidating his own rights, William the Conqueror built a small castle here. Bishop's Stortford has a grammar school, and formerly returned members to Parliament. The church, which dates from the tenth century, is an imposing Perpendicular edifice, and stands on the site of an earlier building. (pp. 23, 24, 79, 92, 112, 132, 144.)

Boxmoor (1127), a small town on the Grand Junction Canal and North Western Railway. A Roman villa was discovered here some years ago. (pp. 132, 133.)

Braughing (930), a village to the north-west of Stortford, on the Cambridge road, with a station on the Great Eastern

Railway, situated in the valley of the Quin. It dates from Saxon times, when it was known as Brooking; and it was granted a market by Stephen. A Roman sarcophagus and many Roman coins have been discovered in the parish. (pp. 22, 113, 136.)

Broxbourne (748) is also a village on the Great Eastern Railway, to the south of Hoddesdon: it contains an almshouse for poor widows founded in the year 1728. The village is intimately connected with the Rye House Plot. (pp. 128, 142.)

Bishop's Stortford, and the River Stort

Buntingford (1272) is a market-town, with almshouses, on a branch of the Great Eastern Railway running northwards from Stortford. It was granted a market by Edward III. (pp. 54, 128.)

Bushey (2838), a parish in the south of the county, separated from Watford about 1166. The village is now the site of the Herkomer Art School. The church was "restored" in 1871, when a late Gothic window was removed.

Bygrave (148), a small market-town a short distance north-east of Baldock.

Cassiobury, a park and mansion at the north-west of Watford which has for many generations been the residence of the Earls of Essex. The present house is modern. (pp. 20, 21, 45, 84, 85, 136.)

Cheshunt (12,292) is a large market-town in the south-eastern corner of the county, nearly north-west of Waltham Abbey, with a station on the Great Eastern and some distance from the town itself. It is celebrated for its nursery gardens, roses being especially cultivated. Within Cheshunt parish is situated Theobald's Park, at one time a royal residence. Cheshunt Park is on the opposite, or north side of the town; and near by are the remains of an old nunnery. (pp. 43, 99, 100, 128.)

Codicote (1145), a small village to the north-west of Welwyn. The church was an ancient one, but a drastic "restoration" in 1853 destroyed much of the evidence of the age of its constituent portions. (pp. 43, 128, 130.)

Elstree (1323), a village on the southern border of the county lying a little west of the Midland Railway, on which it has a station. It is rapidly becoming a suburb of London. (pp. 7, 11, 52, 126, 132.)

Flamstead (1039), a village near the Watling Street to the north of Redbourn. The name is supposed to be a corruption of Verlampstead, the Ver flowing in the valley below the village. The Thomas Saunders almshouses were built in 1669. Beechwood, the seat of the Sebright family is in the parish.

Great Gaddesden (746), a village in Dacorum Hundred to the north of Hemel Hempstead. Gaddesden Place, which was burnt down in 1905 and rebuilt, is the seat of the Halsey family, who possessed the neighbouring "Golden Parsonage" so long ago as 1544. The church probably dates from the twelfth century. (pp. 19, 35, 42.)

The Hadhams—Much Hadham (1199), Little Hadham (655)—two villages lying respectively to the south-west and north-west of Bishop Stortford, and known to have been in existence in the time of the Conqueror. The manor of Hadham

The College Chapel, Haileybury

Hall was granted by the crown to the Bishops of London at the time when the survey recorded in Domesday books was made. (pp. 12, 117.)

Harpenden (4725), a large village or small town on the Midland Railway, almost exactly half-way between St Albans and Luton. During the last twenty years Harpenden ("the valley of nightingales") has nearly doubled in size, and is rapidly increasing. Within the parish is the agricultural experiment-station of Rothamsted; the laboratory being situated on the borders of the village itself. About a mile to the north is Shire-Mere, a small green partly in Hertfordshire and partly in Bedfordshire, and in consequence a favourite site for prize-fights in the old days. Harpenden has branch-lines connected with the Great Northern and the North-Western Railways. With the exception of the more modern tower, the church, which was largely Norman, was pulled down and rebuilt in the sixties. A Norman arch remains in the tower. (pp. 7, 36, 37, 38, 43, 45, 47, 66, 67, 75, 96, 117, 130, 132.)

Hatfield, or King's Hatfield (4330), is a small town on the main line of the Great Northern Railway, chiefly noteworthy on account of its connection with Hatfield House, the seat of the Cecils, Marquises of Salisbury. As mentioned above, Hatfield was at one time a royal palace; but the original building is now used as a stable and riding-school, the present house being of Jacobean date. The residence at Hatfield of Queen Elizabeth is connected with the old palace. Among the features of Hatfield House are the marble hall, its oak-panelled walls hung with tapestry, and its panelled ceiling painted; the grand staircase, hung with portraits; the long gallery, with its armour and pictures; King James's drawing-room, a magnificently decorated apartment; the great dining-room, with a bust of Lord Burleigh; the armoury; and the beautiful chapel, with its exquisite Flemish window and a marble altar-piece. Hatfield is an important railway centre for the county, the Great Northern having branches to Hertford, St Albans, and Harpenden and Luton. Petty sessions are held in the town. In the church are the monuments of the Cecil family, and a statue of the late Lord Salisbury, erected by county sub-scription, stands at the park entrance. (pp. 20, 23, 31, 34, 43, 83, 115, 116, 128, 132, 142.)

Hemel Hempstead (11,264) is an ancient borough and market-town on the western side of the county connected with the main line of the North-Western Railway at Boxmoor, and also served by a branch joining the main line of the Midland at Harpenden. In addition to a mayor, Hempstead has a borough official known as the high bailiff. The town, which is situated in the Gade valley, and formerly returned members of its own to Parliament, is remarkable for the length of its main street—part of which is known as Marlowes. Its market-day is Thursday, and there is an annual wool-sale. Corn and cattle are its chief trade, the straw-plait industry having nearly died out; but near by is Nash Mills, the site of a large paper factory. The church, which stands to the west of the main street, is a fine example of a cruciform twelfth-century parish church; it was commenced about 1140 and finished some 40 years later. There is no evidence of any earlier building on the site. (pp. 43, 47, 101, 102, 122, 131, 132, 133, 138, 147.)

Hertford (9322), although by no means the largest town as regards the number of its population, occupies the first place, as being the county-town, and the only one in Hertfordshire where assizes are held. It is also a market-town and borough (with a mayor and corporation), and formerly returned members of its own to Parliament, although now it is only the centre of a parliamentary division of the county. In addition to the assizes for the whole county, quarter-sessions for the eastern division of Hertfordshire are held in the Shire Hall. The site of Hertford Castle—a building of great antiquity—is now used as the Judges' lodgings in assize time. Hertford has branches of Christ's Hospital, for both boys and girls; and within a short distance is Haileybury College, now a public school, but formerly the training-place for the civil service of the old East India Company. Hertford is served by branches of both the Great Eastern and Great Northern Railways, and has also water communication with London by way of the Lea. It is a centre of the waning malting industry. The old church was burnt down some years ago. (pp. 3, 20, 31, 34, 79, 81, 84, 92, 94, 111, 126, 134, 136, 138, 139, 141, 147.)

Hertingfordbury (733), a village on the railway to the west of Hertford, dating from Norman times. The manor of Roxford was granted by William the Conqueror to Goisfrede de Beck for good service rendered.

Hexton (155), a village in a small parish of Cassio Hundred on the north border of the county jutting into Bedfordshire. Ancient coins have been found in the parish, which includes the old earthwork known as Ravensburgh Castle. Hexton seems to have been granted on two occasions to the monastery of St Alban.

Hitchin (10,072) is one of the most ancient towns in the county, and is now an important railway centre, as it is the starting point of the Royston and Cambridge branch of the Great Northern Railway, on the main line of which the town itself is situated. Hitchin is one of the four parliamentary centres of the county, and is noted for its corn and cattle market, and also as being one of the few places in England where lavender is cultivated for commercial purposes. Hitchin preserves the remnant of an ancient monastery in the almshouses known as the Biggin, and teems with buildings and sites of antiquarian interest. Petty sessions are held in the town. The parish church is one of great beauty and interest, mainly of the Decorated and Perpendicular styles. A picture of the Adoration of the Magi presented in 1774 is believed to be by Rubens. (pp. 14, 68, 69, 74, 105, 106, 108, 120, 128, 136, 141, 142, 145, 149.)

Hoddesdon (4711), an ancient market-town on the eastern border of the county, approached from either the Broxbourne or Rye House stations of the Great Eastern Railway. It is intimately connected with the story of the Rye House Plot (see page 85). It may be mentioned here that the "great bed of Ware" is now preserved at the Rye House.   (pp. 85, 86.)

The Hormeads—Great Hormead (376), Little Hormead (128)—two villages, near the Quin about two miles east of Buntingford, while the latter is about half a mile south of the same.   Both date from the time of the Conqueror.

Ippolits or **Hippolits** (840), a village in the Hundred of Hitchin, dedicated to St Hippolytus patron saint of horses. Travellers used to take their horses to the high altar, where miracles were performed on untamed colts.

Kensworth (516), a small and ancient village in Dacorum Hundred dating from the time of Edward the Confessor, and formerly belonging to St Paul's Cathedral.   The small church dates from about the year 1100, although the tower is later. (pp. 88, 90, 98.)

King's Langley (1579), a village on the North-Western Railway notable as the site of the ancient Tudor Palace of Langley, and of a friary of which portions still remain.   The royal palace and park date at least from 1299.   The friary belonged to the Dominican order.   (pp. 81, 82, 107, 141.)

Layston (983); the original village is now represented only by the ruined church of St Bartholomew, situated a short distance from Buntingford, and of great antiquity.

Letchworth, till recently a very small village on the Great Northern Railway a little north of Hitchin, has now sprung into importance as the site of the "Garden City"; an endeavour to aid in bringing the population back to the land.

North Mimms (1112), a village on the North road, situated some distance to the south-east of St Albans.   A manor of North Mimms was in existence at the Conquest.   The parish includes three large parks, Brookman's, Potterells, and North Mimms. The church, which is rich in monuments, dates from the fourteenth century.

Offley (1001), or Great Offley, which lies on the Bedfordshire border of the county, between Luton and Hitchin, takes its name from Offa II, king of Mercia, who died there in his palace. The church of St Mary Magdalene is built in the Perpendicular style, with an apsidal chancel. Mrs Thrale, the friend of Dr Johnson, lived as a girl at Offley Place. (pp. 14, 105, 128, 141, 147.)

Letchworth, Open Air School

Redbourn (1932), a village on the Chester and Holyhead road, in the valley of the Ver, about four miles north-west of St Albans; it has a station on the Harpenden and Hemel Hempstead branch of the Midland Railway. The manor of Redbourn was given to St Albans' Abbey in the reign of Edward the Confessor. The church, which is some distance from the main street, was dedicated between 1094 and 1109, but the chancel appears to have been rebuilt about 1340. Near Church End are the ancient earthworks known as the Aubreys. (pp. 12, 67, 72, 126.)

Rickmansworth (5627), at the junction of the Colne, Gade, and Chess rivers, is a town in the south-western corner of the county, close to the Bucks and Middlesex borders. It has several ancient almshouses, of which one dates from 1680. Immediately to the south-east is Moor Park, the seat of Lord Ebury, where Lord Anson formerly lived. This once belonged to the abbots of St Albans, but was given by Henry VII to the Earl of Oxford, and in the reign of Henry VIII was the property of Cardinal Wolsey. The present house is of comparatively modern date. The Bury is an excellent specimen of an early seventeenth century mansion. The church appears to have been rebuilt in the fifteenth century. Rickmeresworth was the old name of the town. There are a number of manors in the parish. (pp. 20, 31, 71, 73, 131, 132.)

Royston (3517) is situated on the Icknield Way, actually on the Cambridgeshire border, and is served by a station on the Cambridge branch of the Great Northern. The town, which has a market, stands just at the foot of the chalk downs; it has the honour of giving the name to one of the species, or races, of British birds, to wit, the Royston crow. The church is that of an Augustinian priory now demolished. James I had a hunting seat here. (pp. 32, 33, 45, 93, 128.)

St Albans (16,019), situated about twenty miles north-west of London by rail, enjoys the distinction of being the only town in Hertfordshire entitled to style itself a "city." It is the direct modern successor of the Roman city of Verulamium, lying on the opposite side of the Ver, and itself dates from Saxon times, its ancient monastery having been founded by the Mercian king Offa II in 793, in memory of Alban, the first English Christian martyr. The city has a mayor and corporation, and was formerly a parliamentary borough in its own right, although at the present

11—2

day it forms the centre of an electoral district returning one member to the House of Commons. It is also the centre of the western division of Hertfordshire;—a division corresponding in the main to the old Liberty of St Albans, the area lying within the jurisdiction of the abbot. Quarter-sessions for the western division of the county are held in the Court House, and likewise petty-sessions for the St Albans division of the county, as well as city petty-sessions for St Albans itself. At these last the city magistrates sit; the cases being brought before them by the local police force, which is distinct from that of the county. The city is the see of the bishopric of St Albans, and its crowning glory is its Abbey, now raised to the dignity of a cathedral. Offa's abbey was attacked and plundered by the Danes, and a rebuilding of the monastic church was contemplated by Ealdred, the eighth abbot, who collected building materials from Verulam. The long-deferred work, on a new site, was however not undertaken till the time of Paul of Caen, the first Norman abbot (1077–93). This abbot rebuilt the church and nearly all the monastic buildings with the materials collected by his predecessor; and apparently made a clean sweep of the original structures. Although the fabric appears to have been completed by Abbot Paul, the consecration did not take place till 1115. Between 1195 and 1214 Abbot John de Cella commenced a new west front, but only part of the original design was carried out. In 1257 the eastern end was in a dangerous condition, and the two easternmost bays were pulled down; and eventually a presbytery and a Lady Chapel with vestibule were added. Extensive alterations and rebuilding were carried out previous to 1326, and again between 1335 and 1340. Other works were carried out by John de Wheathampstead between 1451 and 1484, including the rebuilding of St Andrew's chapel. In 1553 the abbey was sold to the Mayor and Burgesses as a parish church, when the Lady Chapel was cut off from the rest of the building by a public passage and used as a grammar school. This passage remained till about

Shrine of St Amphibalus, St Albans' Abbey

1870, when the Lady Chapel was once more rejoined to the main fabric. About this time a restoration of portions of the building was undertaken by a county committee, when the low-pitched roof of the nave was replaced by a high-pitched one on the lines of a much earlier structure. Soon after, the tower was in danger of collapsing, owing to crush in the supporting pillars, and the whole structure had to be shored up previous to underpinning. Finally, the late Lord Grimthorpe undertook the completion of the "restoration," which was carried out in substantial but drastic style. His most notable work comprised the complete rebuilding of the west front in a peculiar style, the repointing of the tower, and the replacing of its brick turrets by stone "pepper-pots."

The clock-tower in the centre of the city, from which the curfew was rung till the sixties, is another interesting building, as is also the old gateway of the monastery, now used as a grammar school. Near by the city are the ruins of Sopwell nunnery. The city has three parishes, those of the Abbey, St Peter, and St Michael, but it is also extending into the parish of Sandridge. On the further side of the Ver is situated St Stephen's. St Albans is rapidly increasing as a residential district, and also as a manufacturing centre, a number of industrial establishments from London having been recently set up in its environs. Straw-plait still remains, however, the chief trade, although the actual plaiting of the straw has been killed by foreign competition. A market is held every Saturday. St Albans has a museum, unfortunately not restricted to local antiquities and natural history objects. There are three railway stations, one on the Midland, the second the terminus of a branch line from the North-Western at Watford, and the third that of a branch of the Great Northern from Hatfield. St Peter's church stands on the site of a Saxon church built in the latter half of the tenth century; this was replaced in less than 200 years by a Norman edifice, remains of which were found during the alterations carried out by the late Lord Grimthorpe. St Michael's church contains Bacon's tomb.

Two notable battles were fought at St Albans during the Wars of the Roses. In 1455 the Yorkists and in 1461 the Lancastrians were victorious. (pp. 14, 18, 35, 57, 59, 60, 61, 71, 72, 75, 80, 81, 82, 84, 94, 98, 102, 103, 104, 105, 107, 123, 124, 126, 128, 130, 137, 138, 140, 141, 142, 149.)

Sawbridgeworth (2085), pronounced Satsworth, is a town on the eastern border of the county, to the south of Bishop's Stortford, with a station on the Great Eastern Railway. It was originally known as Sabricstworth, being the seat of the family of Say, or de Say. It has a history dating from the Conquest. (pp. 113, 132.)

Shenley (1120) a village about four miles to the southward of St Albans; the manor in the time of Stephen belonged to the de Mandeviles, who had also the church.

Standon (1577), a village and manor, with a station on the Great Eastern Railway about midway between Buntingford and Stanstead Abbots. Standon Lordship was the seat of the Lords Aston of Forfar, who inherited it from the Sadler family. The living originally belonged to the Knights Templars.

Stanstead Abbots (1484), now a parish and manor, but formerly a borough, is a village lying east of Hertford, near St Margaret's station on the Great Eastern Railway.

Stevenage (3957), a market-town on the Great North Road and Great Northern Railway, between Welwyn and Hitchin. The town originally stood near to the church of St Nicholas, now half-a-mile distant; but after a disastrous fire, a new settlement sprang up on each side of the North Road, which runs to the south-west of the old church. The fortieth, and last, abbot of St Albans was Richard Boreham de Stevenage, elected in 1538, and dismissed the following year on the dissolution of the monasteries. Elmwood House, now pulled down, was the home of Lucas, the Hertfordshire hermit. (pp. 14, 39, 116, 128, 129.)

Tring (4349) forms the extreme western outpost of Hertford-shire, being situated in the peninsula projecting from this part of the county into the heart of Buckinghamshire. It has a station on the North-Western Railway some considerable distance from the town itself; and of late years has become well-known in the scientific world on account of the private natural history museum established by the Hon. Walter Rothschild in Tring Park, the seat of Lord Rothschild. Tring was formerly one of the centres of the straw-plait industry. (pp. 18, 42, 46, 47, 54, 72, 90, 99, 105.)

Waltham Cross (5291), a town on the Essex border of the county, with a station on the Great Eastern Railway, which takes its name from one of the crosses erected at the resting-places on the funeral route of Queen Eleanor from Grantham. (pp. 124, 125.)

Ware (5573), an ancient town to the north-east of Hertford, situated on the river Lea (which is here navigable), and on a branch of the Great Eastern Railway. Ware, which is associated with the story of "John Gilpin," is the chief centre of the malting industry in the county; the grant of a market was made by King Henry III in the year 1254. (pp. 5, 44, 70, 80.)

Watford (29,327) is by far the largest town in the county, being the only one with a population which exceeded 20,000 at the census of 1901. It is situated in the south-western corner of the county, and is traversed by the Colne; it has a station on the North-Western Railway, from which a branch line runs to St Albans. A market has existed since the time of Henry II, and is stated to have been granted by Henry I. Watford played an important part in Wat Tyler's rising. The Grove, the seat of the Earls of Clarendon, and Cassiobury, that of the Earls of Essex, are situated in the vicinity of the town. Watford is the centre of the West Herts parliamentary division and has numerous mills and factories. The parish church contains some magnificent monuments by Nicholas Stone. (pp. 6, 31, 34, 44, 99, 122, 123, 131, 132, 141.)

Watton or **Watton-at-Stone** (710), a village in the valley of the Beane, near the centre of the county, taking its name from the number of springs in the neighbourhood—*Wat*, in Saxon, signifying a moist place. Watton, which was in existence as a manor in the time of the Conqueror, was the home of the ancient family of Boteler, whose seat was the present Woodhall Park, now the property of the Abel Smith family. Near by is the manor house of Aston Bury, a fine example of a sixteenth century house, also once belonging to the Botelers, with tall, twisted chimneys, a magnificent staircase, and an upper room occupying the whole width of the building. (p. 113.)

Welwyn (1660), a village on the Great Northern Railway between Hatfield and Stevenage. Young, who was born near Bishop's Waltham in Hampshire, became Rector of the place, wrote his *Night Thoughts* here, and is buried in the churchyard. Two centuries ago Welwyn was celebrated for its chalybeate springs. (pp. 62, 128, 149.)

Wheathampstead (2405), a village in the valley of the Lea, between Luton and Hatfield, with a station on the Luton and Dunstable branch of the Great Northern Railway. The parish originally included Harpenden, which was separated about 1860 One of the oldest buildings is Wheathampstead Place, or Place Farm, which dates back to the time of Queen Elizabeth, and has some fine Tudor chimneys; it was formerly the property of the Brockett family, whose monuments are in the church. The church itself, which is a cruciform edifice with a central tower, is dedicated to St Helen, and was judiciously restored in the sixties; the chancel with its three beautiful lancet windows was built about 1230, the tower was rebuilt towards the close of the thirteenth century, and the north transept between 1330–40. The parish includes the manors of Mackery End and Lamer; the latter taking its name from the de la Mare family, by whom it was held in the fourteenth century. Lamer House was rebuilt about 1761. (pp. 38, 67, 90, 97, 101.)

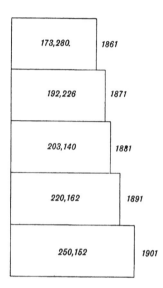

Fig. 1.   Diagram showing the increase in the population
of Hertfordshire from 1861 to 1901

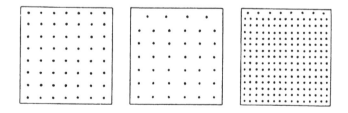

England and Wales 558   Herts 409   Lancashire 2347

Fig. 2.   Comparative density of the population of Hertfordshire
to the sq. mile in 1901.   Each dot represents 10 persons

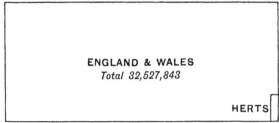

Fig. 3. The population of Herts (258,423) as compared
with that of England and Wales

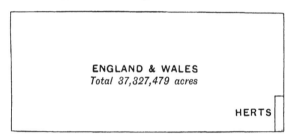

Fig. 4. The area of Herts (404,518 acres) as compared
with that of England and Wales

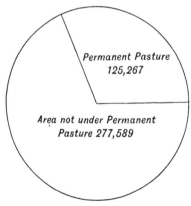

Fig. 5. Proportionate area of Permanent Pasture
to total area of County

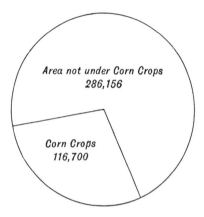

Fig. 6. Proportionate acreage of Corn Crops
to total area of County

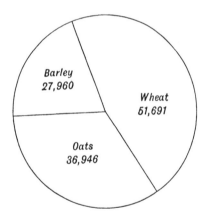

Fig. 7. Proportionate acreage of Oats, Wheat,
and Barley in Herts

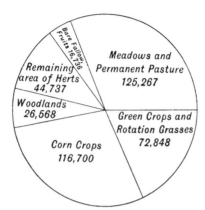

Fig. 8. Proportionate acreage of land under Cultivation
and Not under Cultivation in the County

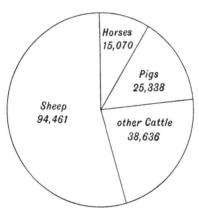

Fig. 9. Comparative numbers of Live Stock in Herts

9 781107 669505